SpringerBriefs in Materials

The SpringerBriefs Series in Materials presents highly relevant, concise monographs on a wide range of topics covering fundamental advances and new applications in the field. Areas of interest include topical information on innovative, structural and functional materials and composites as well as fundamental principles, physical properties, materials theory and design. SpringerBriefs present succinct summaries of cutting-edge research and practical applications across a wide spectrum of fields. Featuring compact volumes of 50 to 125 pages, the series covers a range of content from professional to academic. Typical topics might include

- A timely report of state-of-the art analytical techniques
- A bridge between new research results, as published in journal articles, and a contextual literature review
- A snapshot of a hot or emerging topic
- An in-depth case study or clinical example
- A presentation of core concepts that students must understand in order to make independent contributions Briefs are characterized by fast, global electronic dissemination, standard publishing contracts, standardized manuscript preparation and formatting guidelines, and expedited production schedules.

More information about this series at http://www.springer.com/series/10111

Fabrizio D'Errico

Material Selections by a Hybrid Multi-Criteria Approach

 Springer

Fabrizio D'Errico
Department of Mechanical Engineering
Politecnico di Milano
Milan
Italy

ISSN 2192-1091 ISSN 2192-1105 (electronic)
SpringerBriefs in Materials
ISBN 978-3-319-13029-3 ISBN 978-3-319-13030-9 (eBook)
DOI 10.1007/978-3-319-13030-9

Library of Congress Control Number: 2014960159

Springer Cham Heidelberg New York Dordrecht London

Printed on acid-free paper

Springer is part of Springer Science+Business Media (www.springer.com)

Preface

The engineering point of view is a way of looking at complex problems that are characterized by any mixture of objectives (e.g. material performance, customer receptiveness, monetary, durability, delivery time, positioning strategy on the marketplace, etc.). Typically, by breaking down the problem into more manageable pieces, data and judgments can be rationally and carefully brought to bear on them, after which the pieces are then reassembled in order to present a coherent overall picture to decision makers.

This is also the case when a material selection strategy must be put into practice, for the simple reason that the choice of material can be relevant and have a great impact on the success of a product in the market.

Despite the fact that the literature background of material selection and the specific methods to be used are well-regulated, several objectives often require negotiation, in obedience to different performance targets and different business units and company departments, which may all conflict with each other, thus becoming at best ineffective or, in the worst cases, leading to unsuccessful choices being made.

The scope of this brief is to present a new method that is based on the author and his students' shared experience in applying a structured method that has, as its main goal, the creation of an overall organization in the options, using language and employing a platform that is not restricted to engineers.

The book has been structured into four chapters so that the reader can gradually build up an idea with the given information and it is designed to be used in traditional classrooms, in higher level professional education courses, for self-study, for professional training programs, as well as in many MBA programs focusing on manufacturing management. Thus, it is written for all those people, who aim for a better understanding of how to integrate and speedup the entire product development process from the initial product concept, engineering design phases to design specs, manufacturability and product marketing with optimal choice of materials.

I think a warning is necessary. Depending on your own background, some concepts might appear too specific, and some other topics too specialized, as they could require specific knowledge on materials; as well as that, some other topics might be too boring for you. Even if you have a technical background you might refuse some strange things that you may consider should be a concern for the manager and the marketing department. Do not worry, jump them and go ahead, without anxiety, since that is how we have structured the reading of this book. As in a visiting tour, the first reading could be the first trip for you in a new landscape, so you climb a hill and get a general panoramic view over the valley just underneath you. Maybe you are curious to move toward the valley to enjoy its character and a few of those corners to the fullest, and then hopefully you will join in with some local tastes.

Ultimately, I wish to thank my numerous former students who have participated in this project and its main scope, allowing people with a variety of backgrounds to communicate, negotiate, and decide how to be successful. We hope we have not failed in our job.

Contents

Chapter 1
The Material Selection Strategy

Abstract Chapter 1 outlines the general principles of material selection strategy, and what happens when it is restricted to material specialists instead of being considered as a multifaceted problem managed by teams with multi-layered skills. There is an explanation of the two main approaches used in material selection strategy, the derivative and non-derivative methods, and an outline is provided of the main differences between them. Some available tools are presented.

Introduction

In the spring of 2011, a friend of mine, an engineer who works in a multinational company accepted an invitation to talk to my students who were attending a first-year course in metallurgy and materials. While he does have an engineering background, he is not a metallurgist, and he works in the field of product and market development in a highly diversified international company. He was fervent in talking to future engineers—namely engineering students—discussing competences, skills, the attitudes that people who want to modify and improve the environment and world for their own children have, but he reminded them that the world can be shaped only if everyone concerned reflects on and comes up with choices and decisions. He commented frequently on typical Monday morning business meetings, where, at the beginning at least, people from different company unit areas are sitting around the table sipping from cups of black coffee to try and wake themselves up, everyone with their smartphone in their pocket because they have been told to shut them down so they can concentrate on what is being said. They struggle for a time to keep in stride with the meeting and to participate actively, but technical discussions can often be very difficult for nontechnical people, and in the end, in the stakes between the coffee cup (for concentration) and the smartphone (distraction), it is the cup that loses out to the challenge of the phones, which get turned on again.

That day, an academic and a company manager agreed that it is very important and worrying to consider just how many parts of multifaceted problems relating to product improvement or product development processes (the latter with regard to the most challenging cases) are lost because the multidisciplinary skills of a project team are only potential, however efficiently they may be expressed. This is

© Springer International Publishing Switzerland 2015
F. D'Errico, *Material Selections by a Hybrid Multi-Criteria Approach,*
SpringerBriefs in Materials, DOI 10.1007/978-3-319-13030-9_1

particularly true when a product (re)design process involves the choice of material or material selection process. It is easy enough to leave questions of materials to material specialists or designers, but it is by no means rare that the material potentialities illustrated by such specialists in complicated formulas, data sheets regarding the mechanical properties of on-the-shelf materials, diagrams, and in some other dramatic cases, even by chemistry are finally discarded because the material is too expensive. On the other hand, discarding a choice of the material to be used simply on the basis of its dollar-per-kilogram price is nonsense when different materials with different mass versus volume densities and different resistance limits have to be compared. A major example is what was usually commented on in the past about the use of a magnesium metal-based material as a substitute in the automotive sector. One of the main drawbacks was the higher price of magnesium against aluminum. On the other hand, it was not considered that despite the fact that raw materials are purchased by procurement in dollars per kilogram, the final optimized quantity of material in the end product depends on how much raw material a designer actually needs in order to realize part sections capable of supporting the same external loads in a baseline case.

Sometimes, material selection problems are of greater complexity. This may be the case when the project team is asked not to make a choice from among candidates of the same material class, for example, which is the best contender among alloyed steels for a motor crankshaft. The challenge might sometimes be to create an enlarged point of view. For example, what class of material is the best candidate to give the most powerful and efficient contribution to weight saving in cars, considering that the current snapshot on the average car depicts a percentage lightweight share onboard of about 29 % in 2010 (Mc Kinsey 2010). With regard to the remaining "heavy" share of the vehicle, steels and cast irons dominate.

Reducing the steel and cast iron share in the weight on board an average internal combustion engine motor vehicle is not easy, because they carry load types that make it difficult to find lightweight solutions in the substitution of materials. For this reason, it is really difficult to reduce the weight of a conventional 1400-kg vehicle even by 20–50 kg, a very low percentage. But by decreasing gear components, avoiding crankshafts, and reducing chassis elements, that is, by decreasing the number of components that cannot be made of light materials, the potential lightweight share increases. This is feasible in new battery-charged electric vehicles or fuel cell-powered cars where about 200 kg is a reasonable target.

Material specialists love to claim that materials largely contribute to product improvement and development processes. We like to consider the choice of material as a "quarterback" in several challenges, since such decisions could make a considerable impact on the entire product value chain. Thus, it is interesting to take a brief progressive look along the entire value chain, in order to understand if we are willing to bet on our quarterback, thus deciding to turn off our smartphones and rejoin the meeting we abandoned for a time.

Selection of materials is strictly linked to product manufacturing phases. That is to say, over the life cycle of a product, manufacturing covers the "cradle-to-exit gate" stage, namely acquiring raw materials, shaping them, and (if necessary) carrying out some operations to fit subcomponents and assemble the final product.

Fig. 1.1 Value added is the net output of a sector after adding up all outputs and subtracting intermediate inputs determined by the International Standard Industrial Classification (ISIC). (Source: World Bank national accounts data and OECD National Accounts data files)

Manufacturing is an essential part of economics for many developed countries. It is a fact that a nation that wishes to reach a developed status significantly increases its own proportion of gross domestic product (GDP)-related activities devoted to manufacturing.

A snapshot of the added value of the manufacturing sector per country (Fig. 1.1) depicts the recent situation—an average value of about 12 % of GDP on a world scale, about 20 % if we consider only the developed countries, with peaks of over 40 % in some developing economies. In addition, it is a fact that manufacturing must culminate in products that people need and want, if they are to act in a competitive global marketplace. And it is also a fact that countries whose offer in the manufacturing sector cannot be completely absorbed by internal demand usually create strategies in order not to shrink back to national boundaries. Thanks to the global market, products manufactured in Germany can be absorbed by larger but more competitive markets. Only high-quality products that are reliable, economical, easy to use and manufacture, and can reach the market in a short period time would have a chance for survival. A foreseeable increase in the aggregate demand of high-quality products against price and cost competitive products is an anthropologic fact, rather than an economic fact. As is well outlined by McKinsey Quarterly (2013), the relative role of low-cost products devoted to a vast, developing domestic market has been winding down since countries like China began to get rich. The income of more than half of China's urban households, calculated on a purchasing power-parity basis, will catapult them into the upper middle class—a category that previously barely existed in China. The members of this group are already demanding innovative products that require engineering and manufacturing capabilities that many local producers do not yet adequately possess. The result is that product value chains are becoming more complex, primarily because many new consumers are growing more sophisticated and demanding. The choice of materials is a key issue in the process, which leads to a rise in the complexity of a product value chain.

After the product is on the shelf, namely sold to customers or users (sometimes these can be different subjects), the usage stage starts: Some factors connected with materials can substantially influence consumers' evaluation on usability, quality, and durability. Product development follows, therefore, a two-tier approach: one is externally projected to recognize customer needs and the other is internally focused mostly on the domain of engineering in that it aims to define such key features in a product that can efficiently and powerfully translate customer needs into a success story, possibly surpassing the customers' expectations. It is not rare to find that the choice of materials constitutes the added value of a product. To the touch of the hand, the aluminum used in smartphone cases is perceived as more appealing than polymers. Superior raw material and manufacturing costs for such choices have been powerfully deployed by engineers. Aluminum, they bet, would offer more resistance per gram, and the cover case can be thinner, wider but lighter, thus suitable for a larger and more colorful screen. Apple produces a smartphone 40 % of whose weight is owed to metals (aluminum accounts for 18 %, stainless steel for 16 %) against a very low (1.8 %) weight of plastics, carefully shaping publicity in terms of saying they have created an environment-friendly life cycle for electronic goods mass production. After the usage phase, the choices made on the material to be utilized, decisions that were taken in the design phase, also determine end-of-life management since the product can be disposed of,reused, or in the best cases, recycled. During both the usage and end-of-life phases, there are some negative externalities: The choice of higher-carbon-footprint materials could impact economically on revenues if governments choose to reduce such externalities. As opposed to that, choice of recyclable materials would be a strategic decision to progressively reduce the higher cost of new metal alloys that are expensive today but have high recycling potential. Recycling reduces the impact of the cost of virgin materials.

For example, it is known that the choice of lighter materials to put on board a vehicle fueled by gasoline can contribute to reducing emissions, namely the carbon footprint of the vehicle over its life. A fuel–mass correlation factor (which Ridge (1998) sets to 1.08×10^{-4} kg CO_{2eq}/km kg) can help to point out how the linear function emissions over travel distance can be redrawn by a gentler slope as shown in Fig. 1.2. On the other hand, lighter components made of metallic materials, such as aluminum, magnesium, and titanium, are usually "dirtier" in the extraction and shaping processes, even though the direct emissions from their usage are lower than those of heavier materials. The emissions "stored" in manufacturing have, therefore, to be accounted for in the net balance between the cleaner and dirtier phases when, in order to make a decision, we compare lighter scenarios with those of the baseline.

Minimizing the consumption of energy during manufacture and use, increasing energy consumption from renewable sources, as well as maintaining water and air quality at the highest purity levels are all societal benefits when they are carefully considered in product plans and designs. While in the past relevance of such issues was devoted to sustainability consciousness of material specialists, product development engineers and designers, it has proven that today further efforts put into designing and development of sustainable products may also have positive impact onto competiveness of products.

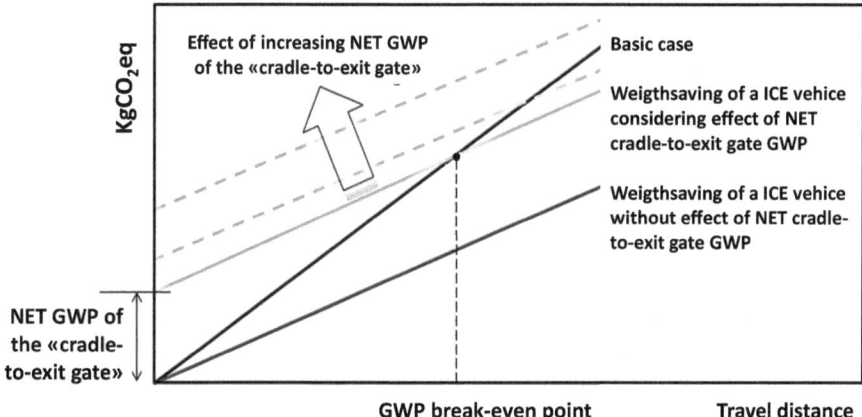

Fig. 1.2 Scheme for the assessment of positive or negative global warming potential (GWP) of lighter automotive component: seeking the GWP break-even point

Careful selection of materials may give control of the quantity and types used in the manufacturing phase, thus leading to a powerful re-think on assembly and how product functions can be maintained and repaired. As consumers attribute importance to a product's usability, among their own interests, an optimized configuration of material-manufacturing-costs is a concurrent added value that consumers recognize in competitive products.

On the market domain, reducing emissions is becoming a further added value in competitiveness. The US government is making important progress toward reducing greenhouse gas (GHG) emissions. A large number of US states and localities are implementing clean energy incentives and clean energy targets—from voluntary emission goals and green building standards to mandatory cap-and-trade laws similar to the carbon taxes of EU Directives, which give polluters a financial incentive to reduce their GHG emissions and encourage the reduction of negative externalities in their own production processes.

Scope of Material Selection

Illustrious (Farag 1989; Dieter 1991; Charles et al. 1997; Lewis 1990; Ashby 1999) and more recent milestone literature (Ashby and Brechet 2004) in the field of the selection of materials approaches the problem in terms of strategy. Thus, by introducing the term strategy, the problem of material selection is enlarged to define not just a method but plans and actions that will on a long-term basis have a substantial impact on the success of a product on the market, and consequently on the success of firms against competitive forces (Porter 1980, 2008).

What we can state at the beginning of the story is that the selection of materials is a complex and multifaceted challenge that requires teams with multidisciplinary skills. Such skills shall be completed by seeking people who may remind engineers that something which is the best is not always something that can be sold, if the customer does not perceive the difference. As well as having in the team people capable to argue for considering cheapest solution for material chosen today would make projected long-life product costs rise. And, lastly, it is a matter of fact that teams who do innovative things are made of people capable to explain technology in easy words.

Whatever the final objective or objectives of the strategy are, a material selection strategy is performed proceeding with three main consequential tasks as follows:

1. The translation of customer or user needs (i.e., external analysis) for the product, influenced by material features into technical and nontechnical requirements (i.e., internal analysis);
2. On the basis of a technical and nontechnical set of requirements, namely the *material key-features inventory*, as developed in (1), the formulation of performance *metrics* to measure how well a material matches a set of requirements;
3. A search procedure, namely a structured material selection method, in order to: (a) explore a solution space, (b) identify materials that meet the constraints, and (c) rank them by their ability to meet the requirements.

A selection strategy works by defining how it is possible to convert a set of inputs, which we call the *requirements* of the product, into a set of outputs. In other words, an effective material selection strategy shall pursue solutions capable to put into a shape customer requirements and contemporarily giving perception to users their expectations have been finally satisfied.

What does the triple-tier approach consist of in practice? That is, what do the formulation of constraints, the formulation of metrics, and the defining of search procedures mean, and what role do they play? Given that there are two different types of methodologies that we can follow, firstly it is convenient to fix the key concepts, and then, as soon as we have a hold of the main ideas, we should consider refining such concepts by discussing these two macroapproaches.

Let us consider an example.

When we go to the bike shop and look around to buy our new bike, firstly we are conscious about our budget. Our budget is obviously related to the price, but not exclusively. We consider several aspects at the same time in order to judge whether our main needs could be satisfied by the bikes we are "screening" in front of our eyes, e.g., whether they give us the right kind of feeling that they will be light and robust enough for the main type of riding we do. Weight and frame rigidity is often an issue for many bikers. A bike manufacturer knows that the targeted market segments are sensitive to this aspect, but customers are sufficiently specialized to understand that lightness must not exclude robustness and durability. All these external requirements are input requirements collected from the market (often referred as the "*Voice of Customer*" or *VOC*) but they need to be translated into technical functions. The translation process from the external or customer needs to the internal or designer requirements will be discussed in depth in Chap. 2. Here, it is enough to focus on the general process. For example, while the customer thinks about a lightweight bike, the engineer thinks of the density of construction material. Furthermore, when a biker is willing to pay $ 700 for a lightweight bike but he wants to put it under pressure by riding it off track, he will probably expect the bike to be robust enough and not to break all of a sudden. The engineers therefore think about the toughness, resistance, or resilience capability of the frame material. This process is what we call in material selection strategy the translation of needs (or external) requirements into technical (or internal) requirements. If you acute reader have gone beyond the work of the writer, you may already have identified the following crucial key point: Once engineers have translated the customer needs into measureable parameters, namely, once they have defined the *metrics*, how can they set the optimal values to target? Actually, this is the crucial question that most literature in the selection of materials tries to answer by introducing rigorous analytic methodologies, as we will discuss in the next section. At this moment, consider that engineers can define the main target for the metrics more or less precisely, namely for the technical requirements, or key factors, that can impact positively on the customer needs. Finally, the introduction of internal metrics and targets allows engineers to *rank and screen* various candidates in order then to select the one that matches the targets and respects technical and nontechnical constraints. One of the major nontechnical constraints that bike engineers fix is a target price that is not lower than marginal cost, as they are not indifferent to the budget constraints of the buyer. Other constraints are directly expressed in the technical language of engineers, e.g., how they translate the safe behavior required and expected of the bike frame when users ride down the hill. Engineers fix a minimum value of toughness resistance for the materials to screen, since they acknowledge that materials with lower values could possibly fail, leading to sudden and dangerous breakages at high speed.

What happens in every material strategy we plan is that several requirements that aim to answer customer needs will compete for fixed resources, such as the cost budgeted, the manufacturing and assembly time, and weight reduction, which contends against the strength of materials.

We have to make choices in a scarcity environment. In other words, we need to seek the optimal solution as the most efficient and powerful compromise between all the aspects we define as important, not just a high-performance solution regarding a few of these aspects. It is easy to find everyday examples of how people make purposeful choices when they are confronted with a scarcity of time or resources. If students spend all their time on metallurgy and materials, they can get full scores in the final exam, but that might mean getting a zero in machine construction. The choice made by efficient students is to balance out their time in order to get a decent grade in both subjects. This is what an economist knows well: The total amount of resources in an economy—workers, land, machinery, factories—is limited. A choice must be made.

The best way to keep in mind all the features product materials should satisfy is to consider them as various aspects of a multifaceted problem. Such a problem will be analyzed by multidisciplinary experts from the marketing area capable of hearing the customer's voice, from the procurement area for negotiating the costs of materials, from the research and development area for discussing performance targets and solutions, from the production area to mitigate impracticable solutions for the installed plant, from finance, to evaluate investments, e.g., in a new plant in order to satisfy technological breakthroughs that need introduction.

But all those people around the table need to speak a comprehensible and common language so that they can negotiate what is important and what is not, so they can decide what they are willing to discuss, make arrangements, and move toward a midway solution.

Material Selection Methodologies: An Overview

Material selection methodologies in engineering design are devoted to giving support to decisions often made in uncertainty and in the awareness that multiple conflicting criteria need to be managed and satisfied. Material selection methodologies are thus constructed to seek out the optimum choice of materials throughout a combination of certain key factors (controlled independent variables or design parameters) which permit the obtainment of a product characterized by a defined number of desired properties (dependent variables, quality response characteristics, functional requirements). The general cognitive scheme is well known in much literature in various fields (from economics to medicine and engineering) dedicated to making decisions in multiobjective optimization[1] (or MOO) problems.

A decision-making problem is defined as *multiobjective* if its solution consists of identifying multiple objectives that somehow have to be combined in order to yield one final solution. It is not likely that a single solution that simultaneously maximizes all the objectives (as we discuss in depth in Chap. 2) exists. As in the simple example of the bike development engineer, in order to realize a competitive product, engineers are generally asked to pursue some macroobjectives such as: (a) minimizing mass; (b) minimizing volume, so that less material is used (sometimes space is precious, as in mobile phones, portable computers, etc.); (c) minimizing material and manufacturing costs and increasing productivity; (d) minimizing the environmental impact, which usually involves positive impact in reduction of energy per kilogram of product, thereby reducing sources used and direct production costs.

The choice of a specific material among varieties that can simultaneously minimize mass, maximize stiffness, and minimize cost in light of all the given design constraints means that it is necessary to study the problems by considering that:

---

[1] Multiobjective optimization (also known multicriteria optimization) developed in the broad area of multiple criteria decision making, also known as multiple-criteria decision analysis (MCDA). MCDA is a subdiscipline of operations research methodology founded in the UK during World War II explicitly to consider multiple criteria in decision making in highly complex environments like the military. Generalizing the operational research and derivative methods further developed concerns with mathematical optimization problems involving more than one objective function to be optimized simultaneously, so that decision makers could pursue the best choice.

a. The properties of different materials are in conflict with each other and that we cannot simultaneously optimize all of them;
b. Beyond a certain trade-off, we arrive at the improvement of one feature of interest at the expense of worsening another; and
c. The search for balance is reached by optimizing the resolution of the trade-offs that exist between features and also by excluding solutions that are not respectful of constraints.

The above boundaries are typical for a multiobjective optimization problem. We discuss in the following paragraphs the two main approaches or methods that are applied to material selection strategy problems, dividing them into two macrogroups: (a) the nonderivative, or *implicit* methods, and (b) the derivative, or *explicit*, methods.

Derivative Method of Material Selection Strategy

A derivative method, or *explicit* method, for material selection starts from translating product requirements and needs into an objective function. For example, a reasonable objective function for the bike is minimizing the weight while still making it possible to charge the external load as the biker uses the pedals, inducing torsion and a bending state of stress in the frame tubes. The elastic deflection of the tubes caused by the biker pedaling has to be minimized, since the higher the elastic deflection during pedaling, the lower the energy transferred to gear and wheel. This is a further constraint with regard to the supporting load.

An explicit method thus starts from the objective function and expresses itself in terms of material features—e.g., density, elasticity module—and other parameters that depend on design.

Generalizing, such methods consist of writing out the *explicit* form of the objective function in such a way as to show how it depends on the material variables that can be used to rank the candidate materials.

In order to reach a better understanding of the process, let us consider the following typical classroom case of material screening among candidates.

The case consists of selecting the best candidate material to reduce weight on board a vehicle by substituting panels made of steel (as a baseline scenario), with magnesium alloy or aluminum alloy. The objective function therefore is to select out of the three screened materials the one that offers the lightest solution alternative to a real-case scenario that is defined by the following design constraints:

• Steel pan is the baseline scenario, and it accounts for about 30 kg (for the sake of simplicity, the case study considers a single large pan, while in reality 30 kg is the total weight for around five pans) onboard and have length L, width a, and thickness b as specified in Table 1.1;
• The load on the baseline pan is fixed and there is a momentum M_f that bends the pan.

Table 1.1 Comparison of weight saved on board when either conventional wrought magnesium or wrought aluminum is used to substitute a steel pan loaded in bending mode to external momentum M_f

Features	Magnesium alloy (type ZW30)	Aluminum alloy (type AA7050)	Steel (type AISI 4140)
Density (kg/dm³)	1.81	2.74	7.87
Resistance limit (MPa)	125	220	450
Substitution factor (kg alternative/kg baseline)	0.44	0.49	1.00
a (dm)	19.49	19.49	19.49
b (dm)	0.019	0.014	0.010
L (dm)	19.49	19.49	19.49
Volume (dm³)	7.21	5.43	3.80
Total weight onboard (kg)	13.1	14.9	29.9
Weight saved (kg)	16.9	15.0	0

Instead we conceive the redesign phase as something that will take advantage of a number of free options (namely the "free-variables" as engineers put it) left for the new component, such as:

- The thickness of the pan b, which is, however, limited to twice the baseline pan;
- The material, which means the material resistance limit.

Two real candidates are therefore considered for substituting steel pan: a wrought magnesium alloy, a commercial ZW30 series; and an aluminum alloy, a commercial AA7075 heat-treated alloy. The two material family groups, magnesium and aluminum alloys, are selected because it is known that potentially these two material classes offer advantages in weight-saving strategies because of their density, i.e., mass per volume, lower than steel. As shown in Table 1.1, the densities of magnesium and aluminum alloys are respectively 22 and 34 % that of steels.

On the other hand, 4140 steel has a higher resistance limit[2] than the two other candidate materials. This implies that we would expect to use a much greater material mass in order to realize new pan sections in either magnesium or aluminum that could support the same momentum M_f, namely the external load, as supported by the baseline steel pan.

On the other hand, due to the lower density, each kilogram of steel is expected to be substituted by a smaller mass of light alloy. This key concept in automotive sector literature that deals with vehicle weight-saving is known as the substitution factor. The meaning is: how many kilograms of a baseline scenario material can be substituted by 1 kg of a lighter material?

---

[2] In the case study, the resistance limit of a material is considered to be the resistance to bending fatigue, thus assuming the pan is loaded by this load type. This would not be true in a real case scenario, since the pan is not highly stressed by fatigue loading. In any case in many classroom cases, we prefer to consider some rearranged situation, reflecting on the simple shape and geometry of a component in a state of stress which would be usual in many other structural parts of the object.

Fig. 1.3 Critical stress to
calculate in order to compare
the resistance of a material in
case of bending load

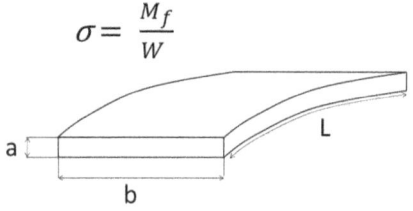

$$\sigma = \frac{M_f}{W}$$

As the reader can guess, it is not correct to calculate the substitution factor as the ratio between the density of the baseline steel and the density of the light material. This procedure does not take into account the lower resistance offered by lighter materials, as commented and shown in Table 1.1.

We need to take a further step.

As stated, regarding design geometry constraints, the width a of a panel is fixed as is the length L of the panel, while the thickness b of the cross section is free. We can reduce the mass by simply reducing the cross section, but there is a constraint: The section area A must be sufficient to carry the bending moment M_f. We can solve this problem in the following way.

As we know (or our engineers must know, if we are not technicians) the state of stress σ induced by the external momentum M_f is easily calculated by the formula in Fig. 1.3 as:

$$\sigma_{appl} = M_f / W \tag{1.1}$$

where W is the moment of resistance to bending. For the study case, it is calculated as:

$$W = \frac{a.b^2}{6} = \frac{a}{a} \cdot \frac{ab^2}{6} = \frac{A^2}{6a} \tag{1.2}$$

Note that we express the W in a more convenient way as the function of the area A and the geometry constraint we have for the design, as the width of the panel a.

The design load constraint that requires the pan to resist safely to the bending moment M_f can therefore be expressed by a relationship that states that the strength of material σ_f, the stress to failure, shall be higher than the maximum stress applied to panel σ_{appl} when it is supporting the external load bending momentum M_f:

$$\sigma_f \geq \sigma_{appl}, \tag{1.3}$$

Combining the design failure constraint in Eq. 1.3 with Eq. 1.1 and 1.2, we can rewrite Eq. 1.3 as follows:

$$\sigma_f \geq \frac{6a \cdot M_f}{A^2}, \tag{1.4}$$

This last relationship means that the panel should be designed by a cross area A capable of keeping the second member below the threshold σ_p, which is the strength limit of the chosen material. Thus, considering we are addressing a minimizing weight problem, the larger the area A is, the heavier the component will be. Thus, Eq. 1.4 will actually be considered with its lower bound value, namely:

$$\sigma_f \geq \frac{6a \cdot M_f}{A^2} \tag{1.5}$$

As our first step consists in calculating the substitution factor for alternative materials, we need to write the ratio between the mass m (obtained multiplying density ρ by section area A—that is expressed by inverting the Eq. 1.5—by length of panel L, in case of constant thickness) of the alternative panel against the mass of the baseline panel, that is:

$$\frac{m_{alternative}}{m_{baseline}} = \frac{\varrho_{alt}}{\varrho_{base}} \cdot \frac{\sigma_{base}^{1/2}}{\sigma_{alt}^{1/2}} \tag{1.6}$$

Table 1.1 gathers the results of the calculation of the mass of alternative panels made in two new materials against baseline steel, considering the constraint of thickness for the alternative light panel that will be limited to twice the baseline steel panel.

  This method is defined as explicit or derivative because we derived the objective (minimum mass among alternatives) by variables that depend in some way on materials and design geometry. Recognizing and applying proper design constraints (maximum thickness of new panel; panel must resist to baseline case load), a general calculation procedure can be defined to screen two alternative candidate materials for reducing the mass of the part. The method used takes into account how the component works and how new materials contribute to the targeted reduction in weight.

Nonderivative Methods for Material Selection Strategy

A non-derivative method or implicit method does not derive from a specific objective function in order to calculate the optimum. In the previous example, the objective remains, but it is not known how the mass of the part can be expressed in terms of the resistance of materials, external load, and part geometry. Instead, the nonderivative methods are also known as "black box methods." The most recognized is the weight-sum method. It consists of:

1. Identifying the key features of a material that can impact the main objectives. For example, in the case of the lightweight panel previously discussed, the mass of the part and the cost of the material are among the desired key features. The choice of material can impact both themes; particularly, the specific strength, i.e., both resistance limit versus density and dollar per kilogram for raw material procurement should be considered;

2. Assigning weight factors to each key factor by creating a qualitative rating (e.g., weak importance = 1, medium importance = 3, high importance = 5); in this step, a quantifying judgment is made on the relative importance of product key features for the product in alignment with customer/user requirements;
3. Quantifying a numerical value that scales the key feature values for candidate materials to a simplified rank (e.g., weak = 1, medium = 3, high = 5). For example, let us consider the three materials for the automotive panel, the 4140 steel, the 7075 aluminum alloy, and the ZW30 magnesium alloy. We deliberate on their specific strength, namely the material limit resistance versus density. A simple calculation leads to 69.1, 80.3, and 57.2 (MPa dm^3 kg^{-1}), respectively, for the magnesium alloy, the aluminum alloy, and the steel. Thus, these three values can be scaled to a simplified rank from 0 to 5 in two main different ways:

   - *Rate by the largest value:* Divide the three values by the largest of the pool and obtain the three normalized values 0.86, 1, and 0.71. Multiplying these normalized values for the maximum simple scale value, namely 5, the final quantifying values on a simplified 0 to 5 rank are therefore 4.30, 5.00, and 3.60;
   - *Rate by the average value:* Calculate for the three values to 69.1, 80.3, and 57.2 MPa dm^3 kg^{-1} the mean value 68.8 MPa dm^3 kg^{-1}. Now consider that on a simply 0 to 5 rank, 3 is the average value. By similitude between average value on a real scale and a simplified scale, the unknown ranked value can be calculated on the simplified scale as: $x = 3 \cdot (specific\ strength)/(average\ specific\ strength)$[3]. Calculation for the three specific strengths of materials leads to 3.01, 3.50, and 2.49.

As the three steps above have been completed, the result can be as shown in Table 1.2.

Referring to Table 1.2, for the three materials' P, the results are 2.1, 5.2, and 5, respectively for the magnesium alloy, the aluminum alloy, and the steel. The material aluminum alloy AA 7075 with the largest value of the sum P is considered as the best for the application. The alternative ranking "rate by average value" would lead to different results as shown in Table 1.3.

This method is sometimes considered inherently unstable and sensitive to alternatives (Ashby 2004). This is clearly shown in Table 1.4, which gathers data for a light (low-density) component that must be strong as well. Data for four possible candidates are shown. The total performance index P is calculated in the third column.

The carbon fiber-reinforced polymer (CFRP) obviously wins, magnesium is second (minor negative value), aluminum third, and finally steel.

However, if we remove CFRP from the selection because, for example, it is judged too expensive, the new selection among three candidate materials would reverse the ranking of magnesium and aluminum.

---

[3] This formula is based on a simple proportion as (*unknown value*): 3 = (*real value*): (*real average value*).

Table 1.2 Example of use of weight factors and the "rate by the largest value" calculation for scaling real values to simplified rank 0 to 5

Key factors	Importance (weight factor)	Key features	Magnesium alloy (ZW30)	Aluminum alloy (AA 7075)	Steel (AISI 4140)
Resistance limit	5	Strength	1.4	2.4	5.0
Reduce weight	−5	Density	1.2	1.7	5.0
Increase stiffness	1	Young module	1.1	1.7	5.0
		P	2.1	*5.2*	5.0

Table 1.3 Example of use of weight factors and the "rate by the average value" calculation for scaling real values to simplified rank 0 to 5

Key factors	Importance (weight factor)	Key features	Magnesium alloy (ZW30)	Aluminum alloy (AA 7075)	Steel (AISI 4140)
Resistance limit	5	Strength	1.5	2.5	5.1
Reduce weight	−5	Density	1.3	2.0	5.7
Increase stiffness	1	Young module	1.2	1.9	5.8
		P	1.8	*4.4*	2.8

Table 1.4 Example of comparison of "rate by the average value" ranking obtained by omitting and including the CFRP material

Materials	Density (kg/dm³) (weight factor=−5)	Strength (MPa) (weight factor=+5)	P (incl. CFRP)	P (excl. CFRP)
Carbon fiber-reinforced polymer (CFRP)	1.7	700	10.64	18.97
Magnesium alloy (type ZW30)	1.8	125	−*1.94*	−0.45
Aluminum alloys (type 7075)	2.7	220	−2.21	*0.41*
Steel (AISI 4140)	7.8	450	−10.36	−5.00

On the other hand, one major advantage of such a method is the possibility of taking into account various aspects that are not exclusively related to material properties.

For example, it is possible to include the evaluation of production costs as being composed of the cost of materials, machining cost, and surface protection. This leads to an extended analysis.

An instance of this is given in Table 1.5 that completes the analysis in Table 1.2 with the calculation of the global performance index P, which includes the relative influence of product costs. As shown in Table 1.5, when increase of a feature, like cost, impacts with a negative effect, this is accounted by a negative value for the weight factor.

Table 1.5 Example shown in Table 1.2 with the P calculation that encloses further nontechnical features

Key factors	Importance (weight factor)	Key features	Magnesium alloy (ZW30)	Aluminum alloy (AA 7075)	Steel (AISI 4140)
Resistance limit	5	Strength	1.4	2.4	5.0
Reduce weight	−5	Density	1.2	1.7	5.0
Increase stiffness	1	Young module	1.1	1.7	5.0
Product costs	−2	Material costs	5.0	4.4	1.0
		Machining operation cost	4.5	4.8	5.0
		Protective coating/ treatment	5.0	5.0	1.3
		P	−26.9	−23.2	−9.5

This feature makes the nonderivative methods particularly simple and quick to use by a multidisciplinary team, where engineers join with nonengineers to discuss a strategy for the success of the product on the market.

The "Free-Search" Approach by Ashby

Classified as a derivative method for material selection strategy, the Ashby[4] approach is recognized as the most generalized, efficient, and relatively user-friendly method with which to perform quantitative analysis for material selection in engineering cases. In the following, the key steps of the Ashby approach are outlined in order to introduce readers with a nontechnical background[5] to its peculiarities, so that they can acquire sufficient awareness on the matter.

As already illustrated in a previous section, derivative methods carry out quantitative analysis that begins by translating the product requirements and needs into an objective function. More precisely, the aim of the analysis consists in introducing a disciplined approach to a selection problem by identifying its distinguishing key features: what function the component will have, what constraints to take into

---

[4] Member of the Cambridge University Engineering Department where he holds the post of Royal Society research professor, M. F. Ashby is the academic who has, more than others, devoted his research to material selection strategies. Briefly described here, his approach is today recognized by material specialists as the most brilliant and powerful explicit method for engineering material selection.

[5] For readers with a technical background, readers who are not necessarily material specialists but who wish to go into depth in the matter, we suggest: Ashby, M.F. and Johnson, K. Materials and Design, the Art and Science of Materials Selection in Product Design Butterworth Heinemann, Oxford, 2002.

account, what objectives to target, and which free variables to steer through in order
to accommodate optimized solutions. Firstly, the analysis process starts from a consideration of the component's function. Many simple engineering functions can be
described in single words or short phrases, unless we need to explain the function
in detail. For example, in designing a new lightweight panel for the exterior of a vehicle, the objective function in common language would be defined as: "Find me a
material and a proper thickness for a panel of length L and width a (refer to Fig. 1.3)
to support a bending load momentum M_f safely and to make it as light as possible."
In engineering words, this means: As the geometry constraints of the panel, namely
the surface dimensions, have been fixed by the design (the pan has primarily to fit
the body chassis), the optimal choice of material will be the one that satisfies the
major objectives—to resist an external load and keep the pan as light as possible.
Keeping in mind that the thickness of the pan is a free variable, the screening process can actually count on two types of free variables: some pertaining to material
features and only one—the thickness a—to geometry.

   As in any derivative method, here again we have to start firstly from an equation
describing our objective for the panel in Fig. 1.3, namely to minimize its mass. As
introduced by Eq. 1.1, we can express the mass as reported here below, which we
repeat for the sake of convenience:

$$m = \rho \cdot A \cdot L \tag{1.7}$$

It is the equation that targets our objective, so we call it the objective function of the
problem. In the previous section, we finally expressed Eq. 1.1 by design constraints
and material free variables in the form:

$$\sigma_f = \frac{6a \cdot M_f}{A^2} \tag{1.8}$$

Thus, we start from this point and we express Eq. 1.1 as the function of material free
variables, strength ρ_ρ and density ρ, and two assigned design constraints, the momentum M_f and the panel width a. This can be obtained by simply turning Eq. 1.8
into:

$$A = \left(\frac{6a \cdot M_f}{\sigma_f} \right)^{\frac{1}{2}} \tag{1.9}$$

In this way, the optimization function originally written in Eq. 1.1 can finally be
expressed by combining it with the Eq. 1.10. It now looks like this:

$$m = \rho \cdot A \cdot L = [M_f] \cdot [6a \cdot L] \cdot \left[\frac{\varrho}{\sigma_f^{1/2}} \right] \tag{1.10}$$

Look at the objective function, in this new form.

It is an expression of what was stated above: Find a material, that is, act in choosing a specific material that can reduce the mass of the panel while respecting design constraints like the force applied, M_f, and geometry constraints, the width a and length L, by screening materials that have as low a ratio as possible between density ρ and the square root of the strength of the material $\sigma_f^{1/2}$. For Ashby, the objective function can conveniently be written in three parentheses: The first parenthesis contains the load constraints for the case, namely the function the component will have in carrying out its work; the second parenthesis contains the specified and assigned geometry, and the last one the material free-variables that influence the objective function for the study case.

$$
m = \begin{bmatrix} Functional \\ requirements, \\ F \end{bmatrix} \begin{bmatrix} Geometric \\ parameters, \\ G \end{bmatrix} \begin{bmatrix} Material \\ Properties, \\ M \end{bmatrix} \tag{1.11}
$$

And now note the procedure we applied.

The width a and the length L of the panel are specified by design, but we are free to choose the cross-sectional area since the thickness b is a free variable. The objective is to minimize the mass of the panel, m. We thus write an equation for m (cf. Eq. 1.1), which is the objective function we want to minimize. But there is a constraint: The panel must carry the bending load M_f without yielding in bending. Use this constraint (cf. Eq. 1.3) to eliminate the free variable A and read off the combination of material properties in the last parenthesis to be minimized (cf. Eq. 1.10).

Ashby calls the term in the final parenthesis, i.e., the one dependent on material properties, as the material index I of the problem that has to be minimized. Anyway, since it is most usual to think of a material index in a form by which when a maximum is sought, an optimum condition is achieved, Ashby prefers to invert the material properties in (Eq. 1.10) and define the material index I to maximize as:

$$
I = \frac{\sigma_f^{1/2}}{\varrho} \tag{1.12}
$$

One enormous advantage introduced by Ashby's approach is that the three groups of parameters in Eq. 1.11 are multiplied together and can thus be considered separable. This fact implies firstly that the optimum subset of materials can be identified without solving the complete design problem—or even knowing all the details of F and G—and secondly that the overall performance is always maximized by maximizing the material indexes[6].

---

[6] Different combinations of function, objective, and constraint lead in different engineering cases to different material indexes. Although for the sake of brevity we applied the method to a specific case study (a panel that needs to be as light as possible while it carries a bending moment), the Ashby derivative method is general. Therefore, the material performance index I is characteristic

Ranking Materials by Material Indices

Returning to Table 1.1, we worked out a comparison of alternative solutions and baseline scenarios using a very general derivative method to calculate the material substitution factor; thus, we recalculated a mass of alternative scenarios against the baseline scenario. Assessing the final weight saved onboard by alternative solutions (refer to last row of Table 1.1), we stated that magnesium is the optimal solution for the case study. We actually ranked different materials to assess which is the best solution for our constraint problem.

The Ashby method does the same, but in a more efficient way. It is not necessary to calculate material substitution factors, material by material. As stated in Eq. 1.10, or in its general form in Eq. 1.11, the optimal solution is the one characterized by the highest material index I calculated for the specific case.

Such indices, each associated with maximizing some aspect of performance and providing criteria that respect the design constrains assigned, permit a quick ranking of materials in terms of their ability to perform well in the given application. Therefore, what we did at this point was to screen candidates that are capable of doing the job, rank them, and identify those among them that will function in the best possible way, as shown in Table 1.6.

But there is a but in everything! The above method really is powerful since it allows for the screening of various candidates using an absolute quantifying approach. The but is, how can we select such candidates? In the case study we illustrated, the specialists, who had sufficient confidence with lightweight design, were aware of specialized literature indications and knew a lot about the basic features of different classes of materials, selected two possible candidates, the magnesium and aluminum alloys. This is the great advantage of having brilliant students. You might also be lucky enough to have accomplished material specialists and design engineers in your company. But Ashby wanted his method to be an "open" approach that could be followed by nonspecialists as well. If I succeed, he probably thought, such a method would be used both by people who are not experts in materials (which is the typical approach of a good teacher) and also by those who are competent and want to screen among further unexplored solutions (which is typical of someone who wants to facilitate innovation). To solve this problem, Ashby introduced the method by using indices of materials on material charts, as described below.

---

of the proper combination, and thus of the function that the component performs, but it can be calculated in a wide range of problems. Some problems are more complex than the study case we use, and also the equations to derive and to elaborate in the general procedure here described can be of rather greater complexity, even for specialists. Fortunately a wide category of engineering functional problems have been studied by Ashby as master cases. Since it is outside the scope of this book, we suggest that readers who need to select the correct material performance index for their specific case should refer to a fuller catalogue of indices that Ashby has provided in Appendixes in M.F. Ashby, Materials Selection in Mechanical Design, Third Edition, Butterworth Heinemann, Oxford, 2005.

Table 1.6 Comparative analysis by material performance index

Features	Magnesium alloy (type ZW30)	Aluminum alloy (type AA7050)	Steel (type AISI 4140)
Density (kg/dm³)	1.81	2.74	7.87
Resistance limit (MPa)	125	220	450
$I = \sigma_f^{1/2}/\rho$	*6.18*	5.41	2.70
Total weight onboard (kg)[a]	13.09	14.94	30.1
Weight saved (kg)	16.91	15.06	0
	56.36%	50.21%	0.00%

[a] To calculate, refer Eq. 1.10

The Free-Search Strategy: The Material Indexes on Charts

Figure 1.4 shows a strength σ plotted against a density ρ, on log scales. The material indices can be plotted onto the figure using the following procedure[7].

The material index $\sigma_f^{1/2}/\rho = C$ can be written by taking logarithms of the first and second members:

$$\log \sigma = 2 \log \rho + 2 \log C \qquad (1.13)$$

Equation 1.13 is a family of straight parallel lines of slope 2 on the plot of log (σ) against log (ρ), and each line corresponds to a certain value of the constant C. Figure 1.4 shows the 2-slope guideline and the three parallel lines that intercept the three points in the chart identified by the coordinates (125 MPa, 1.81 kg/dm³), (220 MPa, 2.74 kg/dm³), and (450 MPa, 7.85 kg/dm³). They are the three points in the plot corresponding respectively to the magnesium ZW30 type, the aluminum AA7050 type, and the AISI 4140 steel. Each line corresponds to three different values of the $\sigma_f^{1/2}/\rho$ index. It is now easy to screen by plotting the subset materials that optimally maximize performance for the case study[8]. All the materials intercepted by a line of constant $\sigma^{1/2}/\rho$ are of equal performance as a light and safe panel. Those above the line are better, those below, worse, since the former accounts for a higher material performance index and the latter accounts for a lower one (cf. Fig. 1.4).

Note the big advantage of this generalized procedure over more common derivative methods. As stated, the procedure to rank out candidate materials is quick and robust. We can refer to the vast number of resolved cases in component optimization (Ashby 2002), thus applying the proper material performance index to our case. If we have already selected potential candidates, by just comparing the material performance indexes we can rank out the best material for the case study among those selected. But it is possible to widen the selection of candidates starting from

---

[7] For the complete procedure, refer to: M.F. Ashby, Materials Selection in Mechanical Design, Third Edition, Butterworth Heinemann, Oxford, 2005.

[8] For in-depth studies of other specific cases that Ashby illustrates for varying loading geometries and objective functions, refer to the charts with proper straight parallel guidelines (Ashby 2005).

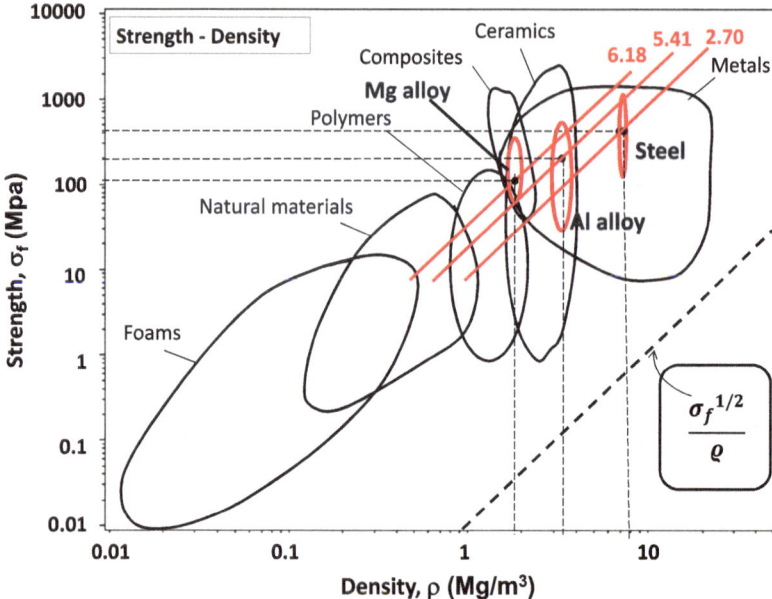

Fig. 1.4 The material chart plots one property against another, in this example, the strength versus density and the mapping out area occupied by each material class, and the sub-fields occupied by individual materials. By coordinates in terms of strength and density, three points have been plotted inside each class of three materials. The dotted straight line shown in the right-bottom corner of the plot has a slope 2 on the log scale diagram, and it is a guideline for drawing any parallel straight line with constant slope 2, that means same value of material performance index $s^{1/2}/r$. The higher in the plot is the straight line having same slope (i.e. slope 2 in the plotted example), the higher is the performance material index investigated. This means that, straight lines that are closer to left-upper corner can intercept material families of higher performance, for the specific solution sought. The plot depicts that, by moving from the baseline solution, i.e., the steel 4140 with performance index 2.70, the more efficient solutions for the specific problem can be "freely" sought by aluminum alloys and magnesium alloys, since both of them have higher performance indexes than steels

the baseline scenario, namely a specific point in the plot and then tracing the line of the proper well-defined slope to explore which materials are possible further candidates, considering the area above the line. This approach follows what we call a free-search procedure. Generally, such a wide screening out has to be conducted carefully, taking into account other limitations and drawbacks of the candidate materials that are not directly pointed out by the specific chart. For example, from the Fig. 1.4 plot, if we also do free a search, we could end up considering ceramics as the optimal material for producing a thin pan for an automobile. Obviously, brittleness, cost, and manufacturing issues have to be kept in mind to restrict our free search to suitable materials, as for example, to CFRP or hybrid materials with metallic foams.

The Expert Survey Strategy, an Implicit Method

In selecting a material for a product or a component, the primary concern of engineers is to match material properties to the functional requirements of the component. Experts must know what material-related variables can significantly influence product function, including the type of material, material toughness, hardness, and fatigue resistance. The type of material used for a component, in turn, determines the manufacturing process, as well as all manufacturing process dimensions, such as machinability, formability, weldability, and assemblability. Depending on the specific manufacturing process involved in component fabrication, one or more process variables need to be tested for component and product functionality to be optimal.

These variables may include cutting speed and feed, the depth of the cut, the temperature, the presence or absence of lubricants, the duration of machining, the rate of cooling/ heating, current density and voltage, and the type and amount of solvent used.

The methodology focuses on the development of guidelines based on determining the relationships between product functionality criteria and design and manufacturing variables that are impacted by the choice of material.

Questionnaires can be used to guide the selection of materials, and processes are outlined by documenting the ways in which experts do their work. Information gathered in this way is called "capture from expert," and such a method is based on the arrangement of a subset of specific questions in order that the expert can provide specific answers. It is an implicit method since it is not based on a rigorous process that expresses general relationships between the features of materials and their performance, quantifying such relationships using certain values, scoring some candidates on the basis of this relationship and ranking out others. And it is a method that strongly depends on having access to competent experts. Since experts, however, tend not to agree unanimously in their answers, guidelines would help in making sure the job can be done more efficiently. We will focus on an illustration of the general methodology to be adopted.

The overall scope of the methodology consists in checking with the experts which key features of a material can ensure product functionality by controlling the design and manufacturing variables that greatly impact on the features of the product. Secondly, once such a subset of material key features has been defined, candidate materials can be scored on the basis of suitability. A matrix scoring model is useful in situations where a number of options are available, and the very best must be chosen. Table 1.7 shows how the matrix scoring model works.

The scoring scale used in Table 1.7 ranges from 1 (poor) to 10 (excellent) and is somewhat arbitrary, since a 5-point scale, a 7-point scale, or the one with a greater or a smaller number of gradations can be used. A larger scale with more gradations increases the sensitivity of the evaluation process. The weights chosen for different criteria indicate the relative importance of criteria in comparison one with the other (one can use a 10-point total or a 100-point total). The final score per each option to

Table 1.7 Example of a matrix score for the assessment of options on general criteria relevant for product functionality

General criteria	Weight (w)	Candidate A	Candidate B	Candidate C
Performance	3	3	2	2
Reliability	4	3	5	7
Quality	2	4	3	5
Manufacturability	2	3	4	8
Environmental friendliness	3	2	4	5

be ranked is calculated by the sum of the score per each criteria (5 in the example of Table 1.7) multiplied by the weight assigned for each criteria category (3 for performance, 4 for reliability, 2 for quality, etc.).

The expert fills out the matrix by assigning a weight per category and scoring material options. Thus, the final results are strictly dependent on his knowledge of the subject and the quality of the information he is equipped with concerning the response of each option to the assessment criteria. A good general practice used to reduce the nonobjectivity and variability of such an assessment consists in breaking down each of the five (or more, if need be) general criteria into sublevels. In order to consider a broad and adequate set of selecting criteria, it is useful if the experts' survey can evaluate candidate materials on the basis of their own relative impact on product functionality, that is on the capability of the product to do something in a safe, reliable, user-friendly, and a high-quality manner, taking into account product manufacturability and the environment friendliness of product life.

The assumption is that the proper choice of a material can positively impact each criterion; thus, the expert should be able to highlight the important key features of the materials that can control overall product functionality.

Table 1.8 illustrates a sample list—within each criterion—of the important key factors that can be effective as regards product functionality. Once a subset of subcriteria for the enhancement of product functionality has been outlined, the expert can proceed by surveying the candidate materials on their relative ability to influence each feature of product functionality.

An example of the final matrix score is shown in Table 1.9. It concerns a preliminary quick survey for a case study conducted to single out the most suitable class of material for manufacturing a crankshaft for a commercial vehicle.

The objective was to identify what class of material and what hardening process to use in order to mitigate cases of failure in a crankshaft produced in series. The options screened were (a) carburizing low-alloyed steels, (b) nitriding low-alloyed steels, and (c) induction-hardened carbon steel.

One main drawback of this method is that it implies implicit knowledge on the expert's part. For example, the relative score that the expert assigns for type classes of steels in the "surface hardness criteria" depends on the knowledge he or she has acquired in carburizing, nitriding, and surface-hardened steels. In addition to that, he must know how different heat treatments act on the steel part in terms of the high risk of distortion that components can undergo during the heat treatment stages,

Table 1.8 Subset of material key factors for impacting product functionality

Performance	Reliability	Quality	Manufacturability	Environmental Friendliness
It serves effectiveness of product function	It is suitable for reducing number of product parts	It satisfies consumer wants and needs	It helps the assembly process	It is reusable
It serves to product response to operating environments	It is suitable for reducing subcomponent redundancy	It contributes to defining product key characteristics	It helps to speed up the manufacturing process	It is recyclable
It serves to function performing consistency	It is suitable for easy-to-perform product maintainability	It reduces critical manufacturing and assembly issues	It favors standardized design	It is not a toxic or hazardous material
It promotes product weight saving	It is suitable for controlling environmental condition	It is easy to inspect and test	It favors design simplification	It contributes to material consumption reduction over life-product stages
	It allows rapid diagnosis if failure occurs	It works as it is foreseen	It can reduce machining operations	It is low-energy intensive in extraction and manufacturing stages
	It cost-effectively improves safety factors	It lasts a long time	It is easy to handle	It is suitable for reducing number of product parts
	It balances material strength and resilience	It is easy to maintain	It avoids special finishes	It is suitable for reducing fasteners
	It controls geometric variability in life It keeps wearing out at a low level	It is attractive/appreciable for consumers	It allows design based on existing products	It can easily be disassembled for reuse, recycling
	It has repeatable behavior in carrying similar loads	It allows design and process capability to increase	It reduces manufacturing cycle costs	It contributes to reducing overall number of different materials used
		It contributes to product simplicity		
		It contributes to distinguishing product from concurrent designs		

Table 1.9 Example of matrix score result of preliminary quick survey for case study conducted on material classes and relative heat treatments suitable for a commercial vehicle crankshaft

Criterion	Subcriteria	Weight	Carburizing steel, low alloyed	Nitriding steel, low alloyed	Medium carbon steel, surface induction-hardened
Performance	Wear resistance	1	4	5	3
	Fatigue resistance	5	3	4	3
	Resistance to contact fatigue	1	3	5	4
Manufacturability	Reduction in machining/finishing operations	4	3	5	4
	Process cost	5	4	2	5
	Base material cost	5	3	3	4
Reliability	Distortion control	5	3	5	3
	Resistance to quenching cracks	5	4	5	2
	Process repeatability (reducing nonconformities)	5	4	5	2
Environmental friendliness	Material consumption reduction	3	3	5	2
	Energy saved over manufacturing steps	3	2	3	5
		Total	139	148	140

5 very high, *4* high, *3* good, *2* very poor, *1* not influencing

which could compromise the subsequent finishing operation, thus requiring the product to be given a nonconformity classification or even, in the very worst cases, to be discarded completely. Note however that despite the fact that this survey process is not always carried out in a such a formal way, it is usually conducted when a company's R&D service starts the auditing process for the selection of suppliers.

Revolutionary Design and Material Design by Analogy

If you take a trip round Barcelona, Spain, you will marvel at Gaudí's architectural work, which, in a sort of gothic naturalism, disseminates a complex translation of nature's forms into the design of buildings and parks. Turtles, seashells, fruit, trees, and other natural forms were part of the inspiration for Gaudí's work. And it was Gaudí's creative genius that brought new curves and shapes that changed the face of

Fig. 1.5 The treelike column structure, inclined and branchlike trees realized by Barcelona's most famous modernist architect, Antoní Gaudí

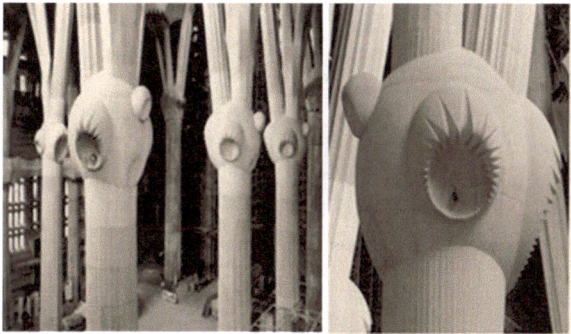

architecture and building technology during the late nineteenth and early twentieth centuries into his the *Temple Expiatori Sagrada Familìa.*

This grandiose church is conceived in an organic style, imitating natural shapes with their abundance of ruled surfaces. He intended the interior to resemble a leafy forest (Fig. 1.5), and in order for that to be so, he searched out new shapes for the inclined columns that emerge like branching trees from a capital that is an element in transition between the trunk and the branches of this "tree." Gaudí was continuously inspired by his research for new shapes and architectural solutions from *analogy.*

The *analogy* is a comparison between the common features of the two elements; it is a "mental telescope" through which it is possible to analyze ideas. When we use analogy to solve a problem, we look at two elements that are not connected to each other; one of these items is part of the problem while the other has a different scope. We find the relation between the two elements, and by making comparison, we derive a new idea. We see people around us every day solving common life problems by analogy. Attorneys are taught to use cases as precedents for constructing and justifying arguments in new cases even in two situations that apparently have nothing to do with each other or have some merely indirect relationship. The fact is we collect a lot of information in a given situation, and we often store all this subliminally in our minds. It sometimes happens that we get an idea which seems to have come from nowhere, but it is probably *reasoning by analogy* that is doing the job for us.

In engineering, analogy often allows us to get to the heart of the problem, thus bypassing many rigorous steps in problem solving.

The term analogy is not so frequently used in the engineering field, since the expression *biologically inspired engineering* or *bionics* is generally preferred. Bionics has been used mostly to describe the systematic application of biological and botanical analogies to solve novel engineering problems. A biological solution to a similar problem is sought. As an example, Velcro was designed using an analogy with plant burrs. Bioinspired materials are rapidly emerging as the most promising areas of research and scholarship that fall within a broader subject domain—the science and engineering of biological materials. Current literature reveals that there

steel base

"scute-shaped" coating

Fig. 1.6 The scute-shaped structure for developing hard ceramic-based layers to support high expansion and contraction cycles

are ever-increasing numbers of national and international scientific and technical conferences, meetings, and workshops on research themes related to bioinspired materials. For example, in a research laboratory, we had to face the problem of protecting the surface of a steel part from the high rate of wear induced by contact with greatly abrasive slurry made of a mixture of liquid fraction and solid particles of aluminum alloy (such a mixture composing a semisolid metal to be injected into thin shape). A protective layer made of ceramic-based coatings could be used to prevent the severe abrasive and chemical wear provoked by aluminum slurry on a steel surface. Unfortunately, the harsh cycles of dilatation and contraction induced on the subsurface steel by repeated heating and cooling-down phases, and a mismatch between the expansion coefficients of steel and the ceramic coating provokes failure in the coating because of brittle fracturing. The need to have such a hard shell to cover the steel surface, but which must also be sufficiently elastic, means that we require a complex material that fits multiple, competing functions. The harder the material is, the lower is the level of elasticity offered. This is a rule of materials.

Looking into nature to locate a similar biological structure able to function in an analogous way, the dorsal (back) convex part of the shell structure of a turtle was noted. The carapace is a dermal shield that is not a unique part. It is composed of the scutes, which are horny plates made of keratin that protect the shell. The scutes are pieces of the shield which can continue to grow during a turtle's life both in thickness and in surface area since they are linked together by ligaments. Combining these features into one material that is composed of small dense cells of ceramic coating inside a metallic frame network as a support is believed to be a strategy enabling the fabrication of complex artificial materials that capture some of the unique features of rigid but deformable biological structures (Fig. 1.6).

The ability to create such complex structures in a reproducible and controlled manner is the scope of this recent branch of material design and selection for a new generation of smart, functional materials and devices.

Databases and Expert Systems Software: Computer-Aided Material Selection Tools

If materials can be classified into some sort of hierarchy[9] capable of helping the designer to navigate through the different possibilities and come to a decision, processes for determining a comparative analysis among them have to affront the management of large quantities of information. As we have introduced it, material selection also relies on the analysis of performance requirements prior to the combination of the constituent elements to create an integrated design. This can happen intuitively, based on the amalgamation of data and information coming from past experience, or systematically through selective analysis. As we have learned, in a systematic analysis that is the main scope of either explicit or implicit methods, these data are formally ranked and assessed to allow for comparative evaluation in relation to the combinations of sets. Explicit and implicit methods differ only on how to perform such a comparative ranking. Whichever approach you choose, the ability to select materials consists in the accurate translation of the design requirements into specific materials and processes. The development of a number of software programs has so far focused on the creation of screening algorithms capable of collating databases and optimizing material selection processes using a series of selectors supported as far as possible by information on manufacturing processes and links to manufacturers (see Table 1.10).

Among implicit methods (namely the black box approaches), *genetic algorithms* (see Table 1.10)-based software has advantages in that its evolutionary capacity means that, rather than revealing single solutions, it offers many novel alternatives beyond the scope of any human-driven research capacity.

Genetic algorithms are computerized search and optimization methods that work in a very similar way to the principles of natural evolution. They refer to models investigated by John Henry Holland in the early 1970s. Based on Darwin's survival-of-the fittest principles, an intelligent search procedure finds the best and fittest design solutions, which are otherwise difficult to find using alternative techniques. They are a stochastic method of search, often applied to optimization or learning—not just a random search—that use an evolutionary analogy, the "survival of fittest" of a set, which is also called "population." *Population* is constructed in the form of binary strings as if they were "chromosomes" that build up the genetic code, thus the name of the method. The string encodes a "candidate solution." For example, "101110" is a possible 6-bit string, the "chromosome" in candidate population that represents a possible solution to a problem. Bits or subsets of bits might represent choice of some feature, for example:

---

[9] Engineers and material scientists classify materials quite differently in relation to six broad family groups: metals, polymers, elastomers, ceramics, glasses, and naturals. Unlike architectural classification systems that are determined by how the materials are applied, each of these groupings is determined by certain properties that are common, such as chemical composition, ductility, elasticity, yield strength, tensile strength, compressive strength, and toughness.

Bit position	Meaning
1–2	Steel, aluminum, magnesium, or fiber carbon-reinforced polymer
3–4	Thickness (mm), width (mm)
5	Elastic inflection under load (mm)
6	Open shape (yes, no)

Table 1.10 Software used to select materials

Name of software	Objective of software and evaluation method
CES (Granta Design 2004), available at: www.grantadesign.com/	A number of specialized databases on materials and processes and the possibility of creating new ones. Utilizes free search strategies and screening of the database
Fuzzy Mat (Bassetti 1997)	Uses the same database as CES, multicriteria selection using fuzzy logic
CAMD (Landru 1999)	Uses the same database as CES, expert guide for developing the set of requirements implementation of coupled equation and value analysis. Strategy uses free searching and questionnaire and screening of the database by recursive algorithm
Fuzzy Composite (Pechambert Duratti 1997)	Optimization of composite materials selection and dimensioning of structural sandwiches. Uses genetic algorithms, screening, and mechanics modeling of possible solutions to evaluate selection
Sandwich selector (Lemoine 1998)	Optimization of materials selection and dimensioning for structural sandwiches. Uses genetic algorithms, screening, and mechanics selection
Fuzzy Glass (Bassetti 1997)	Optimization of glass composition for properties and process ability. Database for correlation. Uses simple algorithms coupled with fuzzy logic to evaluate selection
Fuzzy extrude (Heiberg 2002)	Optimization of aluminum extruded alloy selection, including extrudability and shape via expert rules, using questionnaires and screening to evaluate selection
Fuzzy Cast (Bassetti 1999)	Optimization of aluminum cast alloy selection, including hot tearing and mold filling via expert rules, using questionnaires and screening to evaluate selection
STS (Landru 1999)	Selection of surface treatments according to compatibility with base material and required function, using questionnaires and screening to evaluate selection
VCE (Landru 1999)	Identification of value coefficients in design procedure from the existing solutions based on analogy
MAPS(Landru 2000)	Identification of possible applications for a material from the properties/performance profile. Uses free search and screening of the data to evaluate selection (Landrau et al. 2002)
Astek Expert (Lae 2002)	Selection of optimal joining methods from the existing solutions, using analogy and case-based reasoning
CES Aesthetics (Johnson 2002)	Suggestion for industrial design from a database of objects using analogy and case-based reasoning
Failure Expert (Bouget 2002)	A guide to failure analysis and possible solutions from a database of cases, using analogy and case-based reasoning

To test *population*, a "fitness" function for measuring performance toward an objective is specifically developed. Using a simulator or analysis tool, the performance of *population* may be tested for various solutions, so to determine how particular sets of "chromosomes" impact on final performance. The *population* always stores all that the computer has "learned" about the solution, to improve or drive the next chromosome set.

But they have a limitation due to their intrinsic limited interactivity with humans.

This is not the case of the Cambridge Engineering Selector (CES), commercially launched in 2000 by Granta Design, a spinout from the University of Cambridge, founded in 1994 by M. F. Ashby. It is a Web-enabled tool for the selection of materials and design information built up on the basis of Ashby's free-search approach, deploying material indices and objective functions. A database contains records for materials, organized in a hierarchical manner, where property data for a wide variety of materials are collected. The data are interrogated by a search engine that offers user-friendly search interfaces. Algorithm specifically designed on material indices and objective functions allows you to construct the bidimensional map (see Fig. 1.4) for each study case and the subsequent selection of candidate materials by using superimposing performance index line.

New Computer-Aided Approaches for Optimal Geometry in Brief

By Mauro Linari[10]

With the rapid advancement of computing technology and mathematical modeling, computer simulations have become common methods for the design of complex systems. Effective utilization of computational simulations in place of expensive and time-consuming experimental tests enables engineers to achieve better designs with reduced cost and design cycle time. To exercise these models intelligently and eliminate the burden of manual iteration, manipulating inputs, and reviewing outputs, optimization strategies are applied in a simulation-based design environment. These strategies search for designs that minimize or maximize design goals or objectives while satisfying all design requirements or constraints. Commonly, the assignment of one material rather than another determines the introduction of different stiffness parameters and strength limits. Once the mathematical model has been built, with the assistance of a computer the designer can carry out different types of both static and dynamic analysis, concerning both the in-linear and nonlinear behavior of the material. Several responses such as displacements, velocities,

---

[10] Mauro Linari, M.Sc. Mech. Engineering has worked at MSC Software since 1988 and has been involved in support, training, development and services activities, mainly in the aerospace and automotive field industries, as senior project manager. He is an expert in finite element modeling, stress analysis, dynamics, and optimization in the linear and nonlinear field.

accelerations, forces, tensions, deformations, and others are investigated. If there is no satisfactory solution, the designer can act in two ways:

- Locally modifying the geometry of the part in order to ensure that during the subsequent analysis the internal stresses in the critical areas can be lower than allowed maximum values for the material integrity.
- Changing the material in order to use one that permits the raising of the threshold limit for allowable stress.

Both solutions aim to minimize as much as possible the stress values calculated for the regions, thereby reducing the values of critical stress or deformation. This is the step when the shape and material are optimized, which is crucial for the success of the project phase and verification. Shape changes or material modification inevitably involves an impact on production costs. This method has a limit in the strong dependence of the final result on the geometry initially chosen. Sometimes, only a radical change in geometry, which is something very difficult to reach starting from the original geometry, could permit the use of a less durable, but cheaper material. To get over this limitation, new automated methodologies have been emerging in recent decades. Among these, topology optimization is the most commonly used. This is based on the definition of the "design space," that region within which the preliminary shape of the component that is to be built should be determined. An automated algorithm[11] determines the optimal distribution of the material in the given space of the project (in which the real structure to be designed must be contained), loads, and boundary conditions (see Fig. 1.7).

On the basis of the results calculated by the solver and by the minimum threshold value defined by the user, the postprocessor is able to suggest where to remove material inside the design space. Thus, the preliminary shape of the component remains identified. Additional constraints due to manufacturing processes such as extrusion and casting (with one or two dice) or geometric features like their (even cyclic) symmetry can be taken into account by algorithms. In Fig. 1.8, an example of the whole optimization process by topology optimization is shown.

Limits in topological optimization of shape consist of the possibility of creating a component by adding, not only removing, material. This limit has been surpassed by the topometry optimization method, a related approach that allows you to vary the thickness of the elements in the design domain in order to determine the optimal layout of a structure. This is different from conventional sizing optimization in the

---

[11] Referring to texts and specialized articles dealing with topology optimization for more details, what is important to understand in principle is the way in which the software conducts the analysis: it performs several cycles of automated optimization (in the place of the designer) and, on the basis of simple relations, it builds a map over the design space which in all cases shows the value of a parameter calculated for each element of the mesh (variable in the range from 0 to 1) which can represent the higher or lower importance of each element in collaboration with the global behavior of the structure. Therefore, the software assigns a "quality or importance factor" to each element (generically defined as "element density"). Proceeding with the automated optimization process—and depending on the element density factors—the two design variables on which the software operates—Young modulus and mass density—vary on the basis of the elements, assuming values that are lower or at most equal to those initially defined.

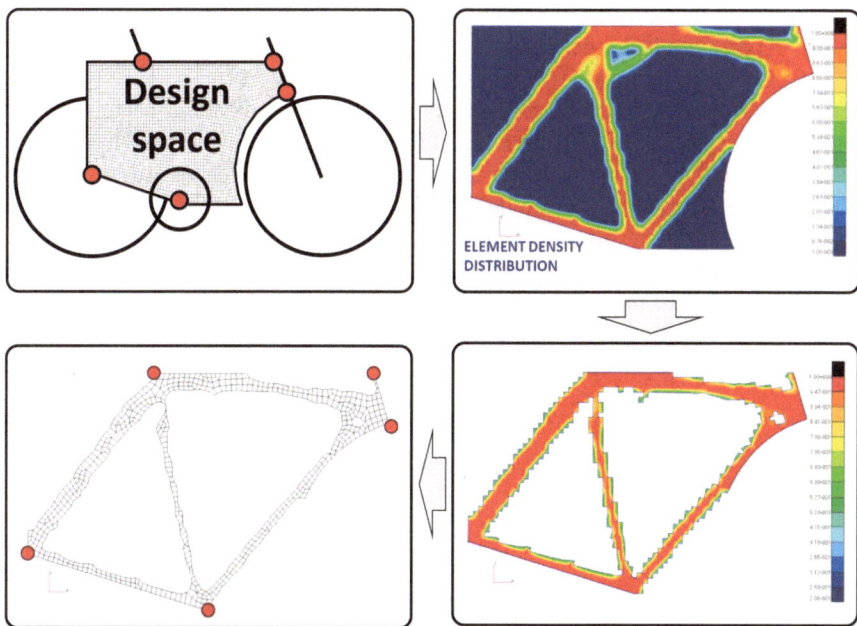

Fig. 1.7 Topology optimization of a bridge: from the design space to the suggested shape for a bridge structure for achieving a smoothed finite element method (FEM) and preliminary geometry (courtesy of MSC Software)

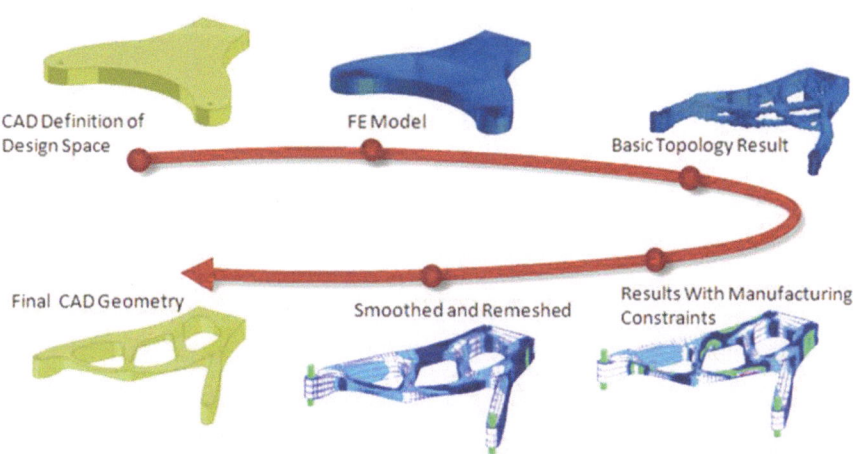

Fig. 1.8 Topology optimization within the design process (courtesy of MSC Software)

sense that in sizing optimization, the thickness of the whole part is changed, but in topometry optimization, thickness is changed "element by element," thus giving greater design freedom. These methods are particularly promising since they can be keyed up for *additive layer manufacturing* approaches, which are rapidly developing because of the impressive advancements in *3D printers* for materials (which might permit the generation of some of the shapes identified without the manufacturing constraints).

In a different way, both topometry and topography optimizations try to solve the same issues regarding the effective use of materials. In fact, while the latter is only able to "flatten" the primary shape provided by the definition of a design space, topometry optimization allows for the redistribution of the mass and stiffness by increasing these characteristics in some areas and reducing them in others. In contrast, topographic optimization changes the structure by modifying the shape (creating beads) according to the geometrical criteria defined by the designer. In practice, these two optimization methods can be considered as the automation, respectively, of "sizing" and "shape" optimization.

Chapter 2
Multi-Objective Optimization in Engineering Design

Abstract **Chapter 2** introduces the reader to multi-objective optimization problems, namely the general approach that is used when complex problems with conflicting objectives are to be faced. This will allow the reader to become confident
with the principle of how an optimal solution is to be sought in a scarcity environment where maximum utility decisions are not configurable with the best materials
as far as their properties are concerned, and with the best compromises needed,
which have to be made so as to compete and succeed on the market. Starting from a
general overview of Quality Function Deployment, QFD - a tool historically developed at Mitsubishi's Kobe - we go on to illustrate how this has been developed for
the precise purpose of screening and assessing engineering materials to maximize
common "utility" as regards "choice in scarcity".

Single-Objective and Multiple Constraint Optimization
by the Explicit Approach

In Chap. 1, we learned the basics of product optimization by carrying out research
on materials using explicit approaches which defines one objective, e.g., "the panel
needs to be light" and one constraint, e.g., "the panel must safely support the external load," and we called a problem formulated in such a way the "strength-limited
design at minimum mass" optimization problem. Furthermore, we learned that out
of all the quantifying—i.e., explicit or derivative—methods, Ashby proposes a
powerful approach that introduces performance indexes for several base cases, thus
providing a rapid ranking and scoring system for possible solutions in order to put
us in a position where we can chose the best one.

But, what happens in the case where you want to select one material among
various candidate materials and you pursue one single-objective—e.g., minimizing
mass—but you have more than one constraint, e.g., a strength limit for the material
and maximum deformation for the panel under load?

Similar cases are frequently found, and here we are dealing with *single-objective
and multiple constraint optimization*. This is usual, e.g., in several automotive components. Brake calipers undergo a continuous improvement as we seek out lightweight metal alloys, and so far, magnesium alloys have proven to be a good choice.

F. D'Errico, *Material Selections by a Hybrid Multi-Criteria Approach,*
SpringerBriefs in Materials, DOI 10.1007/978-3-319-13030-9_2

On the other hand, a designer cannot ignore the fact that such a component also needs to be rigid enough and must keep in mind the maximum volume allowed by the geometry constraints. For this reason, designers consider materials with low Young's Modulus such as magnesium, which are not actually favored to control the entity of deformations due to the stresses induced by external loads.

Let us try to solve this new type of problem by giving a practical example.

Returning to the panel that we have discussed in Chap. 1, we have already completed the task of material selection for the panel; all we did was to apply a quantifying approach based on the Ashby method to the three candidate materials, after which we considered that magnesium was clearly the most suitable material for a single-objective and single-constraint design—in particular, as in this case, we were dealing with a project in which strength was limited by minimum mass (see Table 1.6). Furthermore, as a designer you are asked to pursue mass reduction for a panel that should be sufficiently strong to make cross sections as thin as possible, but which should be, at the same time, sufficiently rigid to sustain deformations under loads. We will deal with the case of a single objective to be optimized—minimizing the mass—but it will now be reconsidered under double-tier constraints:

- Search for a material with sufficient strength limit σ_y. Although the designer likes lighter materials, he has to face the fact that light metal alloys are weaker than heavier ones, and materials that are too weak would force him to increase the thickness of the pan in order to stand up to the external loads, M_f.
- Use a sufficiently rigid material. As stated, the stiffness of the panel is directly influenced by the material's Young's Modulus property. Among the three types considered in Table 1.6, steel has the highest Young's Modulus, i.e., about 210 GPa, and magnesium the lowest, i.e., about 45 GPa.

Fortunately, such a multiple constraint optimization problem can actually be decoupled into two single-objective/single-constraint problems. Referring to Chap. 1, paragraph *"The "free search" by Ashby approach,"* we can write two independent equations and solve each of them by eliminating the dependent variable, i.e., the panel thickness b. The two equations for the optimization problem are written with the purpose of finding a solution suitable for minimizing the mass, by considering the following respectively:

- The minimum material strength σ_y is the unique constraint defined to prevent failures under an external load M_f.
- The material rigidity, the Young's Modulus E, is the unique constraint settled on in order to control panel deformation under an external load M_f.

As stated, thanks to Ashby, we are not expected to be experts in writing such complex equations, much less solving them; as shown in Chap. 1 (cf. Eq. 1.10), it is possible to carry out the material optimization for each design case by isolating the material effect and treating it separately. Great advantages in terms of time saving and reduced efforts are derived from the fact that we can simplify the real case into straightforward base cases that have already been solved[1]. More specifically, among the various cases that have been worked out, we can check the material performance indexes defined for:

- A *strength-limited design* for panels; flat plates that are loaded in bending, where stiffness, length, and width are specified constraints, thickness is a free variable, and minimum mass is the function objective[2]. $I_1 = \sigma f^{1/2} / \rho$ is the suitable material index to fit the purpose[3];
- A *stiffness-limited* design at minimum mass for panels; flat plates loaded in bending, where stiffness, length, and width are specified constraints, thickness is a free variable, and minimum mass is the function objective. The $I_2 = E^{1/3} / \rho$ is the suitable material index to fit the purpose[4].

Let us summarize one more time.

We can simplify a material optimization problem which has a single objective (e.g., choose a material to minimize the mass of panel) with multiple constraints (e.g., the panel must be as thin as possible to save weight but it must be safe in sustaining the load and it must not deform too much) by decoupling it into multiple single-objective/single-constraint subproblems.

In this way, as regards the three candidate materials shown in Table 1.6, we can calculate the two performance material indexes pertaining to each single-objective-constraint problem. Referring again to Eq. 1.10, the mass of the panel is always proportional to the material performance index I, and I varies depending on the load scheme and the constraints that are peculiar to the case. The mass of the panel is therefore proportional to either I_1 or I_2 respectively when either the material strength or the stiffness is the unique constraint. We can therefore calculate the substitution factors for the two subcases as follows:

For the strength-limited design at minimum mass:

$$\frac{m_{1alternative}}{m_{1baseline}} = \frac{I_{1baseline}}{I_{1alternative}} = \left(\frac{\sigma_y^{1/2}}{\varrho}\right)_{baseline} \cdot \left(\frac{\varrho}{\sigma_y^{1/2}}\right)_{alternative} \quad (2.1)$$

And for the stiffness-limited design at minimum mass:

$$\frac{m_{2alternative}}{m_{2baseline}} = \frac{I_{2baseline}}{I_{2alternative}} = \left(\frac{E^{1/3}}{\varrho}\right)_{baseline} \cdot \left(\frac{\varrho}{E^{1/3}}\right)_{alternative} \quad (2.2)$$

Table 2.1 gathers together the final calculations of the mass of the panel in these two cases by varying the material used.

---

[1] Ibid. Ashby, 2005
[2] Ibid. Ashby, 2005
[3] Ibid. Ashby, 2005
[4] Ibid. Ashby, 2005

Table 2.1 Selection of a material for a panel that is light, stiff, and strong

Features	Magnesium alloy (type ZW30)	Aluminum alloy (type AA7050)	Steel (type AISI 4140)
Density (kg/dm³)	1.81	2.74	7.87
E (GPa)	45	70	210
Resistance limit (MPa)	125	220	450
$I_1 = \sigma_f^{1/2} / \rho(MPa^{0.5} dm^3 / kg)$	6.18	5.41	2.70
Substitution factor ($m_{1alternative}/m_{1baseline}$)	0.44	0.50	1.00
$I_2 = E^{1/3} / \rho(MPa^{1/3} dm^3 / kg)$	19.65	15.04	7.55
Substitution factor ($m_{2alternative}/m_{2baseline}$)	0.38	0.50	1
m_1 (kg)	13.09	14.94	30.00
m_2 (kg)	11.53	15.06	30.00
$m = max(m_1, m_2)$ (kg)	13.09	15.06	30.00

Table 2.1 shows that magnesium is still the preferred material when strength-limited and stiffness-limited designs are treated separately. But when we want the panel to meet requirements on both strength, and at the same time stiffness constraints, its mass should be selected as the greater of m_1 and m_2 by using the following criteria[5]:

$$m_{optimal} = max\ (m_1, m_2) \tag{2.3}$$

Here again, magnesium is still the optimal choice since it offers the lowest value of $m_{optimal}$.

Decision-Making in Problems with Multiple Conflicting Objectives Using the Explicit Approach

When you consider real cases concerning authentic products to be optimized in order to compete in the market, your attention as a company manager cannot focus exclusively on performance features. Maximizing performance to surpass competitors does not necessarily mean succeeding in the market. In order to compete and trade in the global market, modern enterprises that are leaders get ahead by surpassing their customer's expectations. Instead of thinking about how to make their own products into technical benchmarks, competitive companies consider any other changes, including material-driven adjustments as opportunity sources for differentiation and for promoting business that can evolve locally—conscious as they are that social aspects and environment preservation are unavoidable today as factors where competition is intense. What emerges is that multiple conflicting objectives

---

[5] Ibid. Ashby, 2005.

have to be taken into account. Saving weight on your car is a frequent goal, as it is on your portable and compact devices, on your sports clothing, and the accessories you buy in order to achieve greater performance. And equally frequent is the fact that such a goal is concurrent with minimizing volume, minimizing the cost, and minimizing the impact of the product on the environment. There are four objectives (i.e., reducing weight, increasing strength, reducing volume, and reducing carbon footprint) that are usually considered, and which are in conflict with one another. As we have seen, engineers seek to *minimize weight* primarily by changing materials and their shape. Usually, however, the lighter the material is, the lower its *resistance strength* becomes, and that forces designers to "add" a volume section at critical points, which in turn is of course a countermeasure that conflicts with *volume reduction*. Finally, light alloys are energy-intensive materials especially during the extraction stage, with the end result of increasing the final *carbon footprint* of the product. We call such cases *multiple conflicting objectives problems,* and they are usually multiple constrained. The solution to them is based on a compromise to be sought: not just by optimizing one single objective, but by researching their optima as well as possible, considering their mutual reliance.

But, what procedure can we follow when we must take decisions on candidate materials so as to arrive at the best compromise? And how can mass, volume, cost, and environmental impact be compared?

Origin and Scope in Brief

We need to look at how a decision model is usually structured in such cases. The basic assumption is that when a multiple conflicting objective arises, decision-making is always restrained by the "decision space": We must make purposeful choices with limited resources. The resources of a material can be, e.g., that it might be lighter, but this is limited by the fact that it cannot be stronger; in addition to that, we look for a material that is both lighter and stronger, but we are limited by the increase in its cost. Economists have developed an operational approach that allows us to assess the public's choice in multiple conflicting objective problems, depending on their own varying preferences. There are different items among a set of alternatives that conflict with each other: You can usually chose to purchase more goods, but you need to devote more day hours to your work to earn the money to purchase these goods; then, the hours you spend at work diminish the leisure time you have to enjoy what you have bought.

Vilfredo Pareto (1848–1923) was probably the first economics researcher whose work might formally be classified as the explicit *multi-objective decision-making* approach. As an economist, he was the first (or at least one of the first) to carry out mathematical studies on the aggregation of conflicting criteria into a single-composite index in order to quantify decision-making assessments. A brief overview of Pareto's basic concepts helps to speed up our comprehension of the explicit method

Table 2.2 Example of *utility* from bananas and kiwis. The numbers inside the box give the utility from consuming the amounts of bananas and the amounts of kiwi shown outside the box

grams of bananas						
1000	1	30	35	39	41	44
800	1	28	33	36	39	41
600	1	26	30	33	35	37
400	1	23	26	29	31	32
200	1	18	21	23	25	26
0	0	0	0	0	0	0
	0	200	400	600	800	1000

grams of kiwis

that we can apply in multiple conflicting objectives with multiple constraint material selection problems.

Consumers have varying tastes and preferences for some goods relative to others. And they normally have limited resources—what the economists call a budget—to satisfy their individual needs and desires. Different preferences imply different choices for the total allocation of purchased goods. One person might prefer pizza and coke for his lunch hour, while another prefers salad and lemonade for his healthy break. And for each of those that like pizza and coke, there is usually a different combination in their weekly consumption of these two things, since one individual thinks that eating "one pizza and drinking three Cokes" per week is better than eating "two pizzas and drinking one Coke." Economists call this "utility"— the fact individuals prefer one item to another. Generally speaking, if activity A is preferred to alternative B, we then say the utility from A is greater than the utility from B. You might imagine you are in an open-air market on a Saturday morning and you are deciding how many kilos of bananas and kiwis to buy with your money for your afternoon frozen shake. If you want, the decision would represent the satisfaction level you reach by mixing—in the way you desire—a different combination of bananas and kiwis. Now observe Table 2.2. It reports the number of grams of bananas listed vertically on the left outside the box, from 0 up to 1000 g, while the number of grams of kiwis is listed horizontally below the box, from 0 to 1000 g. The entries inside the box show the *utility* you get from consuming bananas and kiwis together. Thus, the number of boxes at each intersection of a row and a column means the *utility number* for the consumption and the mixing of a specific combination of bananas and kiwi. For example, if an individual consumes 200 g of bananas and 200 g of kiwis, the number in the box states that the utility is 18. Utility is just a numerical indicator of preferences, but it does not matter what units you use to measure—the fact is that economists have no way of measuring utility. The only rule is that the higher the utility, the stronger the preference. For this reason, it is very helpful to explore combinations in the box and *rank* alternative consumptions. According to Table 2.2, the fans of banana and kiwi-frozen shakes seem to prefer a mixed combination of 800 g of kiwis and 200 g of bananas to a combination of 200 g of bananas and 400 g of kiwis in that the utility of the former (28) is greater than the utility of the latter (21). Thus, by maximizing utility, the consumer is making decisions that

Table 2.3 The numbers inside the box give the total dollar expenditures on different combinations of kiwis and bananas available for $ 3 budget constraint (the beyond-budget combinations are in grey) at $ 0.9/kg unitary price of bananas and $ 3/kg unitary price of kiwis

grams of bananas

	200	400	600	800	1000
1000	$0.9	$2.1	$2.7	$3.3	$3.9
800	$0.7	$1.9	$2.5	$3.1	$3.7
600	$0.5	$1.7	$2.3	$2.9	$3.5
400	$0.4	$1.6	$2.2	$2.8	$3.4
200	$0.2	$1.4	$2.0	$2.6	$3.2
0	$0.6	$1.2	$1.8	$2.4	$3.0

grams of kiwi

Table 2.4 Similar results obtained for the previous Table 2.3 but recalculated by an increased $ 2.7/kg unitary price of bananas and a fixed $ 3/kg unitary price kiwis

grams of bananas

	200	400	600	800	1000
1000	$2.7	$3.9	$4.5	$5.1	$5.7
800	$2.2	$3.4	$4.0	$4.6	$5.2
600	$1.6	$2.8	$3.4	$4.0	$4.6
400	$1.1	$2.3	$2.9	$3.5	$4.1
200	$0.5	$1.7	$2.3	$2.9	$3.5
0	$0.6	$1.2	$1.8	$2.4	$3.0

grams of kiwi

lead to the best outcome from his or her point of view. In this way, utility maximization implements the assumption that people make purposeful choices to increase their satisfaction.

On the other hand, consumers are limited in how much they can spend when they choose between bananas and kiwis and other goods, e.g., suppose the individual allocates a total of $ 3 per week for his favorite fruit. This limit on total spending is called the budget constraint. Tables 2.3 and 2.4 show expenditures on bananas and kiwis for two different situations. In the boxes, all the combinations of bananas and kiwis from Table 2.2 are shown, but some combinations are outside the $ 3 budget constraint—these are in the grey-shaded area. As a result of the varying price per unit of bananas but a fixed budget, some combinations are outside (i.e., much more expensive than) the budget constraint.

Economists prefer to express the data in Table 2.2 by reporting the different values of utility for the varying combinations of possible preference choices in a diagram chart as shown in Fig. 2.1. The axes report the same information about the lines and rows of the matrix in Table 2.2, namely x-axis reports the quantity of kiwis, y-axis the quantity of bananas you chose to mix for your weekly shakes. Note that the chart reports the family curves that interpolate points at the same level of utility (refer once again to the values reported in the cells in Table 2.2 and thus compare the points shown in the chart in Fig. 2.1). And note the procedure used to pass from Table 2.2 to the representation in Fig. 2.1:

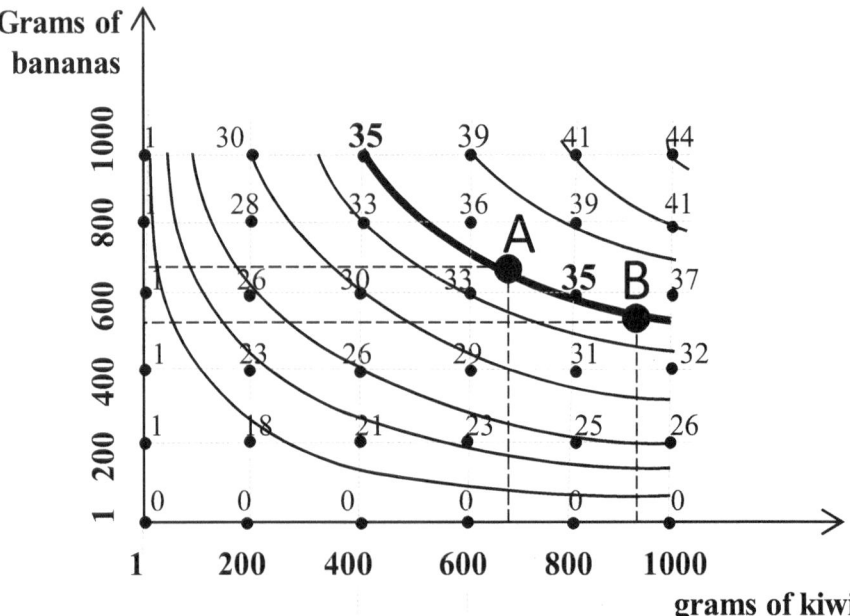

Fig. 2.1 Indifference curves represented on a chart diagram show the utility values of the example in Table 2.2

- satisfaction for varying combination of bananas and kiwis is summarized by fulfilling Table 2.2 that reports the nonmetered values (they are just indicators) of the relative preference for one combination chosen over the other; in economics, such relative preference of choices is called *utility*;
- the values in the cells of Table 2.2 can be transferred to a bidimensional chart;
- curves are added onto chart that interpolate points at the same level of *utility* (i.e., the same level of satisfaction, if you prefer).

We call these curves *indifference curves* because all the points on each curve correspond to different combinations of the quantity of bananas and kiwis (look at the coordinate axes). But each combination of bananas and kiwis can be obtained by moving along a specific curve that leaves the consumer *indifferent* to his choice. For example, the bolded curve in Fig. 2.1 that interpolates points A and B has utility "37"; A and B represent two possible choices for mixing bananas and kiwis that have the same *utility* for the consumer. Thus, he is indifferent as to the choice of either combination A or combination B since he will be equally satisfied.

Let us finally move on to the chart illustrated in Fig. 2.2. It is the same chart we constructed in Fig. 2.1, but with addition of further family lines. They are constructed in this way:

$$p_{banans}\left(\frac{\$}{kg}\right) \cdot q_{bananas}(\text{kg}) + p_{kiwi}\left(\frac{\$}{kg}\right) \cdot q_{kiwi}(\text{kg}) \le budget(\$) \qquad (2.4)$$

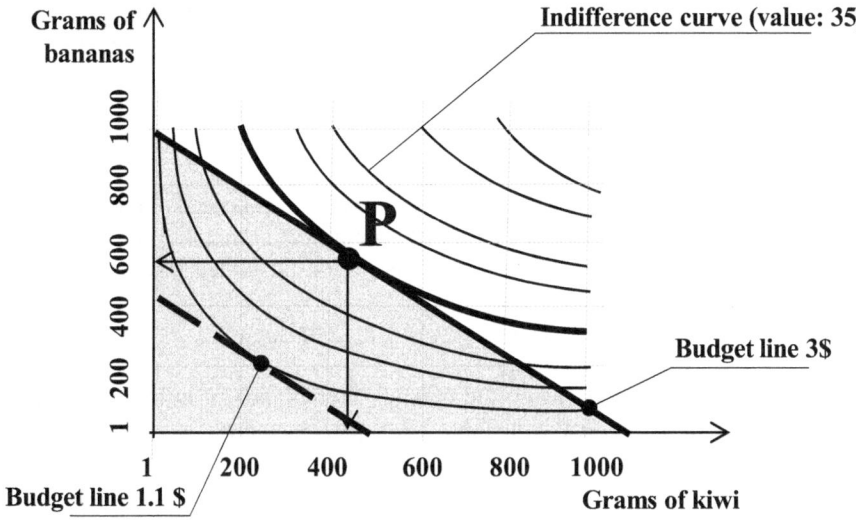

Fig. 2.2 When the *budget line* is tangent to the *indifference curve*, the consumer cannot do any better. Compared to the other points of intersection between the *indifference curves* and the *budget line* "$ 3", this point can maximize utility for the consumer: In fact, the maximum utility (35) is obtained among the *indifference curves* that can match the maximum budget the consumer can spend. The optimal allocation of bananas and kiwis with respect to quantity is therefore defined by this intersection point. Note that if you reduce the consumer's budget—e.g., for $ 1.1 to spend—, the budget line moves on *right-bottom corner*, and thus, the *indifference curve* that allows maximum utility changes accordingly

and we call it a budget-constraint family curve. The meaning is simple: You may buy varying combinations of bananas and kiwis to your own satisfaction, but you are limited by your budget, the $ 3 of total budget you decided, e.g., to allocate to your weekly banana and kiwi-frozen shake. Actually, you can decide to save money from your budget, e.g., spending only $ 1.1. Thus, you might consider the $ 1.1 budget line. But this would mean that you are not maximizing your willingness to allocate up to $ 3 per week to take satisfaction from your frozen shakes. For that reason, the problem of defining the quantity of bananas and kiwis that optimize your budget constraint is solved by selecting any point belonging to the maximum budget line in the chart diagram (see again Fig. 2.2):

$$p_{banans}\left(\frac{\$}{kg}\right) \cdot q_{bananas}(kg) + p_{kiwi}\left(\frac{\$}{kg}\right) \cdot q_{kiwi}(kg) = 3\$ \qquad (2.5)$$

The optimized solution among all the points on the budget line is the one that matches the highest *utility* curve among many. This leads to the definite identification of tangent point P in the chart diagram of Fig. 2.2, a point that corresponds to the quantity of bananas and the quantity of kiwis that satisfy both the *budget constraint* and the *maximum utility objective*.

It is possible to demonstrate, but it is out of our scope[6], that the *utility curves* that are illustrated in Fig. 2.2 can also be analytically represented by the Cobb–Douglas[7] family of curves:

$$U(x, y) = x^a \cdot y^b \tag{2.6}$$

There is no specific reason for using such an analytic form instead of the other one. Such curves were introduced first of all in economics[8] in that they were easy-to-handle interpolating functions. More often, they are used in their logarithmic form:

$$u(x, y) = \log U(x, y) = a \cdot \log x + b \cdot \log y \tag{2.7}$$

The a and b constant depends on the optimal interpolation of utility levels for a specific case. For the case study in Table 2.2, Eq. 2.7 becomes:

$$u(q_{kiwi}, q_{bananas}) = 0.227 \cdot \log q_{kiwi} + 0.319 \cdot \log q_{bananas} \tag{2.8}$$

Furthermore, by solving the generalized analytic expression we achieved in Eq. 2.7 for the utility curves, it is possible to demonstrate[9] that the solution P of Fig. 2.2 has a generalized structure like:

$$\begin{cases} x = \dfrac{a}{a+b} \cdot \dfrac{B}{p_x} \\ y = \dfrac{b}{a+b} \cdot \dfrac{B}{p_y} \end{cases} \tag{2.9}$$

where B is the budget constraint for the budget family of curves $p_x \cdot x + p_y \cdot y = B$ Thus, applying Eq. 2.9 to our frozen-shake problem, we can finally calculate the quantities of bananas and kiwis that solve the problem of spending our \$ 3 weekly budget in a way that will give maximum satisfaction:

---

[6] For the demonstration, you can refer to microeconomics textbooks that deal with the consumer's problem with indifference curves.

[7] In the 1920s the economist Paul Douglas was working on the problem of relating inputs and output at the national aggregate level. A survey by the National Bureau of Economic Research found that during the decade 1909–1918, the share of output paid to labor was fairly constant at about 74%, despite the fact the capital/labor ratio was not constant. He enquired of his friend Charles Cobb, a mathematician, if any particular production function might account for this. This gave birth to the original Cobb–Douglas production function, where K is factor capital, L is labor and A is a constant that represents productivity efficiency (Cobb, C. W. and P. H. Douglas, 1928).

[8] Ibid. Cobb and Douglas, 1928.

[9] Ibid. Cobb and Douglas, 1928.

$$\begin{cases} q_{kiwi}=\dfrac{0.227}{0.227+0.319}\cdot\dfrac{3\$}{2.7\dfrac{\$}{kg}}=0.46\,kg \\[4mm] q_{bananas}=\dfrac{0.319}{0.227+0.319}\cdot\dfrac{3\$}{3\dfrac{\$}{kg}}=0.58\,kg \end{cases} \qquad (2.10)$$

The Cobb–Douglas curves are particularly manageable when the optimization problem regards more than two variables. If we want to add strawberries or any new ingredients to our frozen shake, the optimization problem cannot be illustrated by a diagram chart (in which we cannot deal with more than two variables), but it can easily be settled by the Cobb–Douglas indifference family curve by adding further elements in Eq. 2.11, for example:

$$u=a\cdot\log q_1+b\cdot\log q_2+c\cdot\log q_3+d\cdot\log q_3+\ldots \qquad (2.11)$$

Obviously, the new coefficients c, d, and any others you might consider will be assessed by reevaluating the changes to consumer utility when new ingredients are included, since the original utility values in the matrix of Table 2.2 represent the level of satisfaction from a possible combination of only two ingredients—kiwis and bananas.

Now, we have finally set out some of the basics in the theory of *multi-objective decision making,* how can we use it to the advantage of a multiple objectives material selection problem?

The next step is having an in-depth look through what Ashby proposes.

The Penalty and Trade-Off Curves as a Graphic Method for the Explicit Approach to Material Selection

Returning to our problem in choosing a material for a lightweight but affordable cost panel, as stated in the basic optimization theory, is classified as a *double conflicting objectives* problem—minimizing both the mass and the cost since it is always true that the greater the performance of the material, the higher is its cost. Our scope is to seek out a material among the various candidate materials that will not necessarily be optimal as regards either of the objectives, but which maximizes our *utility.* However, in this case, how can we define *utility,* and what does it mean to pursue "maximum utility"?

Let us proceed step by step. Once again, Table 2.5 summarizes the results of Table 2.1 regarding the mass of the panel for candidate materials against the baseline scenario. As illustrated previously, the mass of the panel can be optimized by

Table 2.5 The same results shown in Table 2.2 to select optima material for a light, stiff, strong panel with further information about the cost of the part made of magnesium, aluminum, or steel

Features	Magnesium alloy (type ZW30)	Aluminum alloy (type AA7050)	Steel (type AISI 4140)
$I_1 = \sigma_f^{1/2}/\rho$	6.18	5.41	2.70
$I_2 = E^{1/3}/\rho$	19.65	15.04	7.55
$m = \max(m_1, m_2)$	13.09	15.06	30.00
Cost ($)	78.55	45.19	36.00

Fig. 2.3 Indifference curves represented on a chart diagram shown for the utility values of the example in Table 2.3

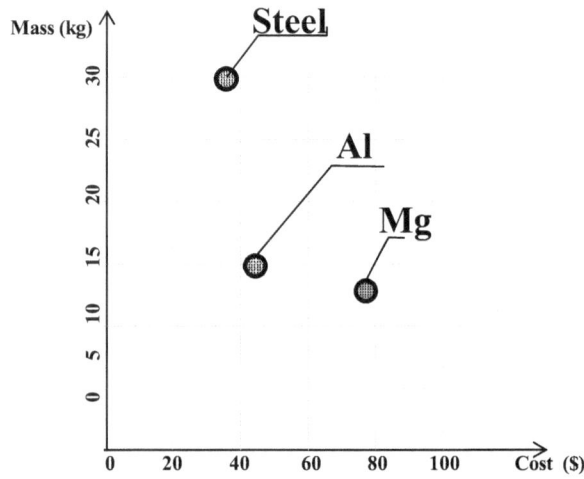

choosing the highest value of mass between the two values obtained for each of the optimization problems we solved separately. Furthermore, Table 2.5 reports the panel cost estimated for each material.

Look now at Fig. 2.3; the diagram shows the three alternative solutions plotting mass against cost. Each bubble describes a possible solution as coming from the results summarized in Table 2.5.

This plot could be sufficient to identify which bubble represents the optimal solution in the case of single-objective problem: If minimizing the mass is the sole objective, magnesium is the final choice. When cost is the dominant criteria, steel is the best solution. But such a plot is ineffective when we consider the optimization of both criteria—mass and cost.

To overcome this obstacle, we can now use some of the powerful tools we acquired when we were looking through the basics of optimization theory. Assume that we know the *utility level* for each bubble as illustrated in Fig. 2.4. Note that unlike goods we purchase for an increase in satisfaction—such as the bananas and the kiwis—in this case, the increase in *utility* for any choices is achieved when both mass and cost decrease. This is the meaning of the arrow in Fig. 2.4: It identifies the direction of the increase in utility for any *indifference curve* that has been plotted.

Fig. 2.4 Indifference curves represented on chart diagram

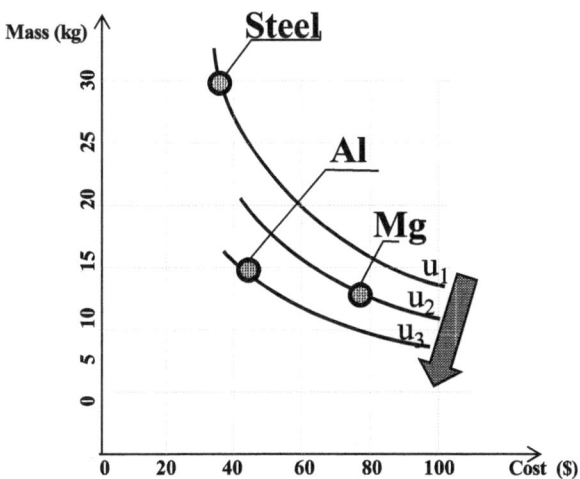

To solve this problem, we employ the chart in Fig. 2.2. The optimum solution P is here graphically sought by checking the tangency point between the budget constraint line and the maximum utility curve. Similarly to the way we proceeded when deciding how many bananas and kiwis to put in our frozen shake in order to maximize our *utility,* the best optimization solution here derives from the tangent point between the minimum utility curve available and a line that corresponds to our constraint.

We now have to solve two kinds of problems:

- The first, how can we actually plot the indifference curves for choices in material on the chart itself?
- Once we have succeeded in solving this, we can face the second problem: How can we define and plot a constraint line that we need in order to seek its tangent point with lowest indifference curve we have drawn for the candidate material?

Actually the first problem is not easy to solve. It is not possible to determine indifference curve for a choice of material; more precisely, though, theoretically speaking, such curves could be analyzed, it would be a complicated procedure, which, moreover, is not effective for us. As stated, our goal is to identify an easy method to use in looking for optimized solutions even in complicated multiple conflicting objective problems. We can proceed in this way:

- locate all the candidates we want to assess by plotting the chart "mass versus cost" (as we did in Fig. 2.4 for the three bubbles)[10];

---

[10] Note that computer assisted (and automated) methods we discussed in the Chapter 1 are a valid support in the phase of generating a possible candidate solution like those represented in the plot of Fig. 13.

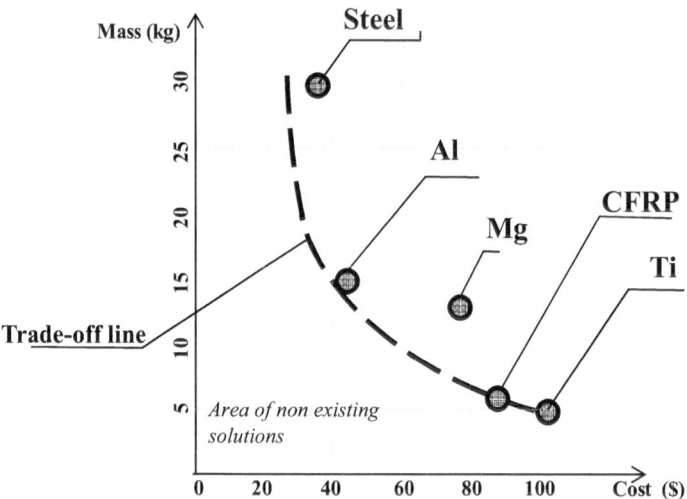

Fig. 2.5 The *trade-off line* helps to single out candidate materials with the best compromise (i.e., maximum *utility*)

- draw the interpolating line that represents the lower bound limit of the existing solution; we call it the *trade-off line*[11];
- select the solutions on or near the *trade-off line* because they offer the best compromise; and
- reject the rest of the candidates (Fig. 2.5).

To select the optimal solution among the candidates that are nearest to the trade-off line, we need to proceed further. We can use two different approaches[12].

The first consists in reformulating one of the two objectives, e.g., the cost, as a fixed constraint: for example, set $ 80 the cost limit that you do not want to surpass. In this way, you can check whether all the solutions that lie on (or nearest to) the *trade-off line* are to be rejected or not, considering this budget limit; in the example of Fig. 2.6, the materials carbon fiber-reinforced polymers and titanium alloys are rejected.

Simple, but it is not actually a true optimization. If we want further to refine the selection among the candidates that are nearest the trade-off line, we need now to introduce an additional element in the plot. We need to initiate something similar to the budget-constraint line, as we did with our budget in the example of our best frozen shake.

We need to introduce the *penalty line*, or *penalty function*[13], as explained in the following. The *penalty function* is the way to aggregate the various objectives, into one single-objective function formulated in such a way that its minimum

---

[11] Ibid. Ashby, 2005.

[12] Ibid. Ashby, 2005.

[13] Ibid. Ashby, 2005.

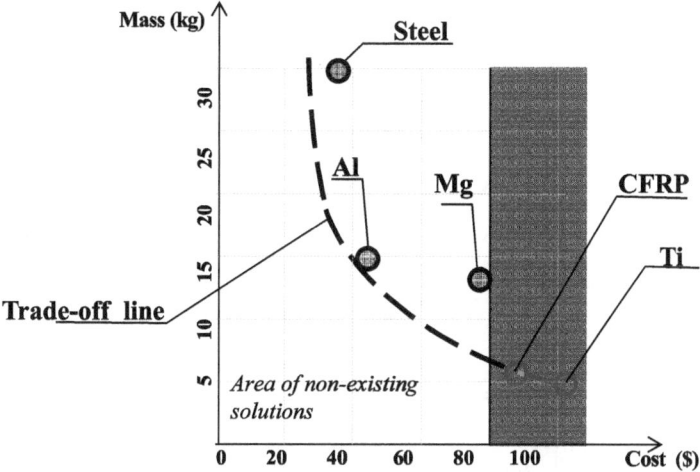

Fig. 2.6 Among the best optimized solutions nearest or aligned with the *trade-off line*, the candidates that overcome a cost limit of $ 80 are eliminated

corresponds to the most favored solution. In our example, the penalty function Z aggregates two objective functions, the mass m and the cost C to be minimized, as follows:

$$Z = \alpha \cdot m + C \tag{2.12}$$

where α is a constant that allows you to convert the mass function expressed in kg into the cost as expressed in a currency, e.g., the \$; thus, the unit measure of α is expressed in \$/kg. Note that in cases where we are dealing with a problem which has four conflicting objectives, we need to search for the best solution in a material that minimizes mass m, volume V, cost C, and the environmental impact of the product E. Then, the penalty function has a structure like:

$$Z = \alpha_1 \cdot m + \alpha_2 \cdot V + \alpha_3 \cdot E + C \tag{2.13}$$

where α_1 units are \$/kg, α_2 units are \$/cm^3, α_3 units are \$/kgCO$_{2eq.}$ In other words, exchange constants measure the increase in penalty for a unit increment in a given performance metric such as mass, volume, environmental impact—all others being held constant. For example, in Eq. 2.13 , if the exchange constant of the mass m is \$ -2/kg, this means that for each kilogram reduced, the cost of the component increases by \$ 2. Similarly, if the exchange constant of the volume V is \$ -10/cm^3, each reduction by 1 cm^3 makes the cost increase by \$ 10. The exchange constant α_i allows us to quantify in a single unit of measurement (the currency), the monetary effect of unitary changes made on optimization variables.

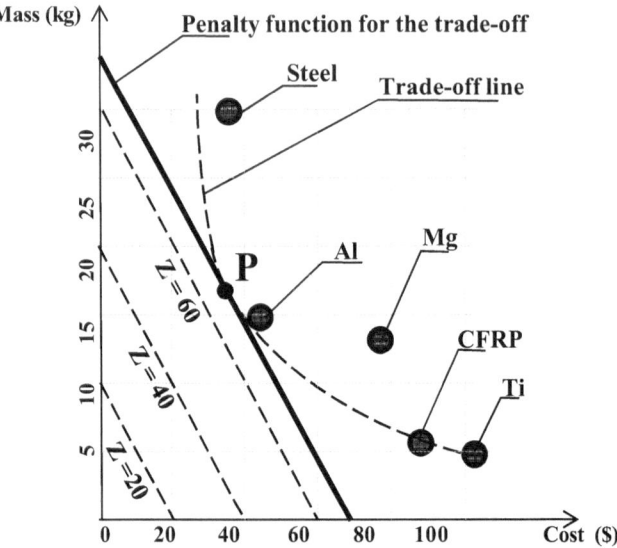

Fig. 2.7 The tangent point between the *trade-off line* and the *penalty function family curve* identifies the zone where the best solutions are located. By plotting it out, it is possible to verify aluminum (*Al*) as the nearest solution to the tangent point, thus the optimal choice for the specific problem

Returning to our example, assuming the exchange constant α is $-2\$/kg$, the double-objectives optimization problem we face is therefore totally defined and can be solved by plotting together (Fig. 2.7):

- the bubbles that locate the candidate solutions by mass and cost;
- the trade-off curve; and
- the penalty function Z.

As stated in the frozen-shake question, such a problem admits a (theoretical) best solution that is represented by the coordinates of the tangent point P. Actually, here we are dealing with real materials and a suitable solution; for this reason, we are not interested in the theoretical optimization problem but in singling out which—among the real, possible solutions represented by the bubbles—is the nearest to what is theoretically the optimal one. From the plot in Fig. 2.8, it is possible to select aluminum as the best candidate material, in that it:

- pursues the objective of minimizing the mass and the cost at a fixed exchange constant value;
- respects the design constraints in terms of maximum deformation and inner stress under load.

Note what has been set out in the first point. On the one hand, it is plain that the answer we get is actually the best solution, since it appears clear from the plot that aluminum is the nearest solution to theoretical optimal solution P. On the other hand, the solution selected depends very much on the exchange constant value that we choose.

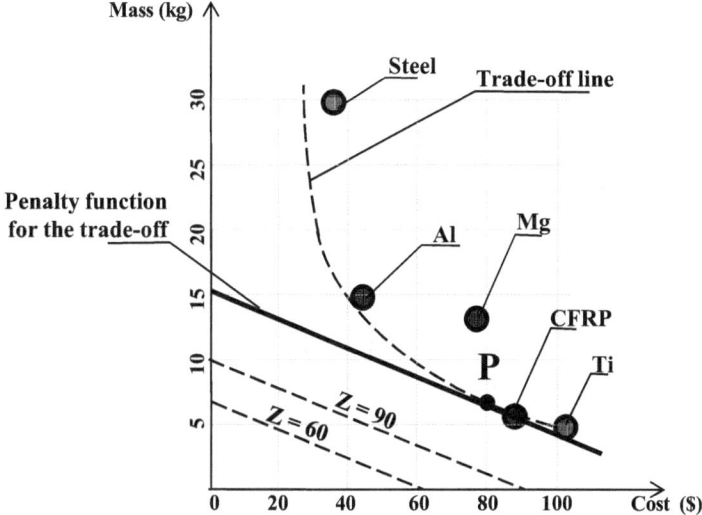

Fig. 2.8 Change in solution driven by the slope of the penalty function: Change in the exchange constant for mass produces a consistent variation in the penalty function slope and therefore the variation in tangent point P

To get a better understanding of this point, assume for your new panel project you are willing to accept that the value of exchange constant α will fall to $ -10/kg. This means that you are willing for your product cost to increases by $ 6 more than the previous case per each kilogram you can save. And what happens if you decide that production cost is so important that minimizing cost is the dominant question in the two objectives? That would mean, e.g., that the exchange constant α is $ -0.5/ kg, and thus, your product cost increases by $ 0.5 per each kilogram you can save on the product. Looking at the graph leads to a foreseeable solution: Only the conventional material, steel, is suitable, and no willingness to change is considered by the company (Fig. 2.9).

As thus emerges, by using the graphic method, it is possible to find more precise solutions to articulated problems that deal with two objectives and multiple constraint problems. Uncertainties are thus restrained at one's own discretion: In a project, you would put more emphasis on either the cost or the performance as a lightweight objective. This makes the solutions vary to three different levels of importance, depending on what is assigned to the exchange constants in the penalty function. Thus, since the values assigned to exchange constants have such a great influence on the final selection of the material, how can we take the right decision on their values?

And, even in cases where we are lucky enough not to have any doubts when deciding such constant values, and in consideration of the fact that the method described above is well settled and defined for no more than 2 conflicting objectives, how can we find a solution to a real problem when we have more than two conflicting criteria to consider?

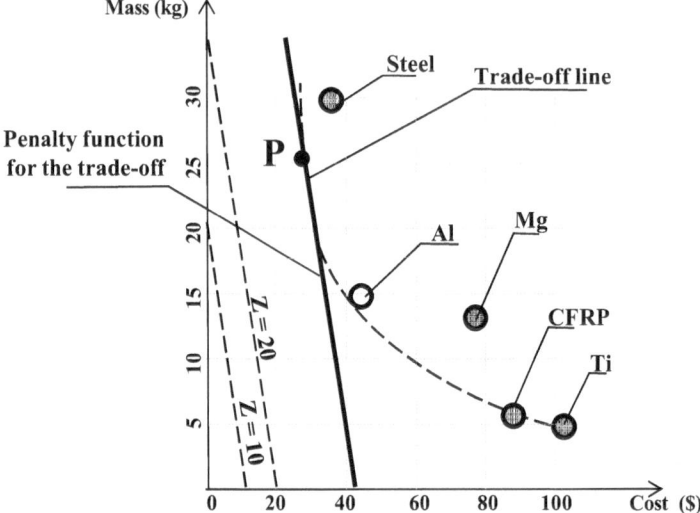

Fig. 2.9 Change in solution driven by slope of penalty function: Lower value of the exchange constant for mass shows that conventional is still the best solution, and no changes in product are considered

We will deal with this type of two-tier problem in the last paragraph of this section, but before that, and in order to make sure that we understand a couple of principles that can be used allowing us conscientiously to make decisions on the method to apply in various situations, it might be convenient to have a brief look at a powerful tool that we refer to as Quality Function Deployment—a product planning tool used most commonly during the design stages of a new product.

The Implicit Approach Using the Quality Function Deployment Matrix

In the late 1960s, at Mitsubishi's Kobe Shipyard, Yoji Akao and Shigeru Mizuno were working on the potential relationships between the customer needs and what technical requirements to target in order to match these needs. Originally developed as a simple matrix that put the customer demands on the vertical axis and the methods with which they would be met on the horizontal axis, further improvements were brought to bear in the early 1970s leading to the definition of a well-structured tool called in English Quality Function Deployment or QFD, which was recognized almost immediately as a major breakthrough. The term Quality Function Deployment is a loose translation from the Japanese name for this methodology, *hin shitsu* (quality), *ki nou* (function), *ten kai* (deployment). In Japanese, "deployment" refers to an extension or broadening of activities and hence "Quality Function Deployment" means that the responsibilities for producing a quality item must be assigned

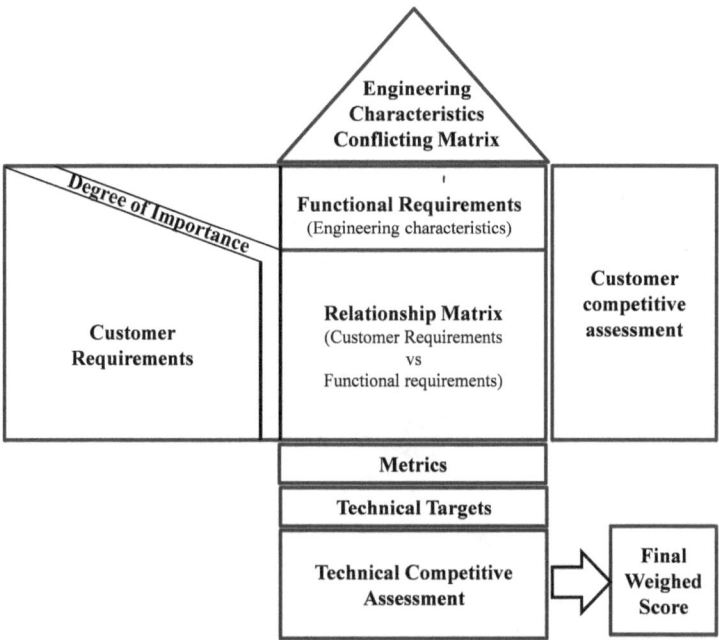

Fig. 2.10 Scheme of the QFD matrix. Because of its shape, it is also usually called the "House of Quality"

to all parts of a corporation. It is sometimes referred to as the most advanced form of Total Quality Control, Japanese style. One very important thing that we need to highlight is that on account of its own structure, QFD is an explicit method for designing a product or service which is based on customer demands and that can involve all members of the organization. A QFD methodology flow is depicted in Fig. 2.10, but the best way to fully comprehend how a method works is obviously to experience it. For this reason, we will now discuss how to construct a QFD matrix for one case study. Firstly, we have to comprehend its general scope: the formation and structuring of a creative engineering process that aims effectively to translate the customer's needs into technical requirements. As such, therefore, QFD is a technique used in order to facilitate the process of converting the customer's requirements, into a product design. If we are engineers and managers, we are usually so close to our product that our level of expectation and our values are far removed from those of the average customer. What frequently happens is that a high-performance product is launched, but it does not meet customer expectations simply because the technical features that characterize the product are not fully perceived by the customer. Or some added features put there to differentiate the product and so to compete in the market are not decisive in targeting the customer's final choice. A general rule used by QFD is that we should not speak for the public, but that we must listen to the "*voice of the customer,*" or *VOC* by gathering the requirements that make their needs more explicit.

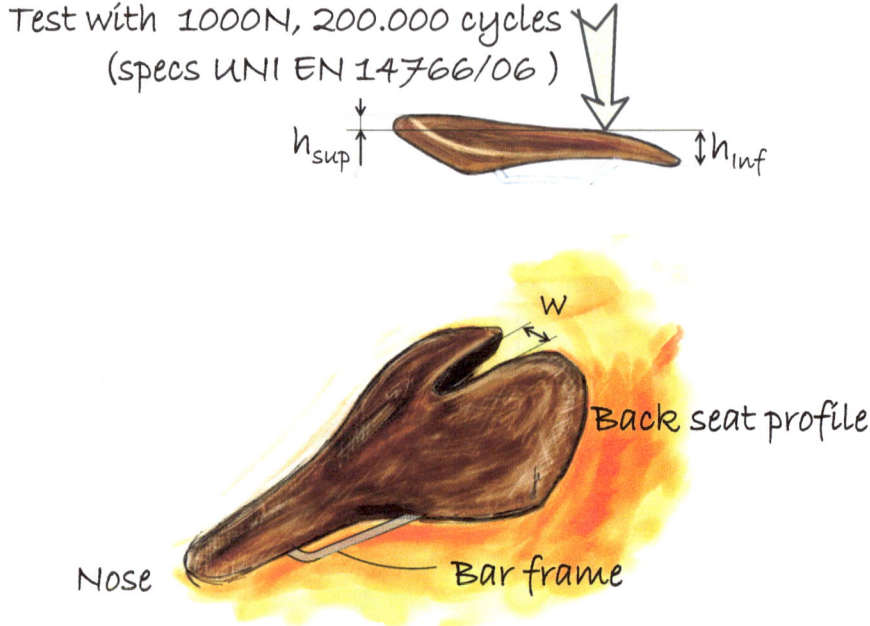

Fig. 2.11 Major features of road-bike saddle

Let us introduce a study case. Assume we are working in a company that produces bike saddles, and we are in a multidisciplinary team concerned with product development. In particular, the team has to seek for a solution to improve the competitiveness of the company's top product, and it has been commissioned to develop new and distinctive features in the product—road-bike saddles (Fig. 2.11).

Note again the process we want to focus on: starting from customers' needs and requirements, which we want to satisfy more completely than our competitors, and find the key technical features that need to be controlled and adjusted in order to influence our customers' evaluation of our saddle in a positive way. Theoretically speaking, such a closed loop is simple:

- gather the customers' needs,
- identify the technical and engineering features that will ensure the customer perceives that our product can satisfy his needs, and
- control the relevant key engineering features in order to surpass those of our competitors.

We have to remind ourselves that a product competes and is successful on the market when such an engineering process finally manages to get on the customer's wavelength and communicate this message: "Buy me! Buy me, since I'm the best!"

Let us have a deeper look. If you talk with expert bikers, they have many comments to make on what a saddle should and should not do. They are your customers, and your scope is now to talk with them and try to understand what they need, what they consider relevant, and what they absolutely hate in a bike saddle.

Table 2.6 Bike saddle example in organizing customer requirements

Primary	Secondary	Tertiary
Performance	It should help to reduce loss of energy by pedaling	It should be light
		It should be rigid
	It should look as if it gives high performance and is aggressive	It should appear aerodynamic in shape
	It should not cause pain even after prolonged sitting	It should be comfortable
		It should feel soft
Cost	It should have low operating costs	It should not require any maintenance by me
		It should be durable
		It should be easy to clean
	It should have a low purchase cost	It should be cheaper
Safety	It fits safety requirements	It should be safe
	It should be safe when it has a failure crash	If it crashes, I must not be injured

We, my students and I, have done the work for you. We looked up specialized books on market research in order to get hold of an in-depth vision on how a customer's needs can reliably be gathered, and you may assume here that a variety of bikers have expressed their requirements and they have been grouped and summarized in Table 2.6. It therefore follows that we can go on to the next step—organizing the customers' data.

- it should be light;
- it should be rigid;
- it should appear aerodynamic in shape;
- it should be comfortable;
- it should feel soft;
- it should not require any maintenance by me;
- it should be durable;
- it should be easy to clean;
- it should be cheap; and
- it must be safe and if it crashes, I must be not injured.

Actually, we need to organize this raw customer data better—into some groups of associated data to simplify using the QFD charts we will be dealing with at a later stage. Because the customer requirements are listed in the language of the customer, the process of organizing the data allows the QFD team to reach a common understanding of what customers want. In organizing the customer requirements, there are three categories into which they can be put. These are as follows:

- *Primary:* The very basic customer wants. At this level all the requirements should give an overall product view. For example, "looks light at first glance," "comfortable to ride," "durable."

- *Secondary*: These requirements are the primary level in more detail and are in fact the headings for groups of tertiary-level requirements.
- *Tertiary*: These are the requirements at their most detailed level.

Entering the Customer Requirements onto the QFD Chart

Once all the requirements have been gathered and organized, they can be entered on the QFD chart on the left-hand side as shown in the diagram in Fig. 2.10.

Establishing Customer Importance Ratings & Customer Competitive Comparisons

After we have gathered the customer requirements, we deal with quantifying:

- The relative importance of each of the characteristics you put in the previous table as they are rated by the customer.
- The score that a customer assigns to our company on each of his/her requirements against the best of our competition.

Once again note that we cannot go in-depth with market research methods, but what it is important to highlight is that whatever survey technique is used to gather customer importance ratings and comparisons with the competition, it is necessary to have both of the above pieces of information. Generally, a 0 to 5 scale is used to simplify. In Fig. 2.12, an example questionnaire is shown, made to collect data on bike saddles.

| | How important this item | | | | | How do you rate our bike saddle on this item | | | | |
	not important	not very imortant	Important	Very important	extremely important	not important	not very imortant	Important	Very important	extremely important
Rate:										
It is light	1	2	3	4	(5)	1	(2)	3	4	5
It is rigid	1	2	3	4	(5)	1	2	3	4	(5)
It appears aerodynamic in shape	1	2	3	(4)	5	1	2	3	(4)	5
It is comfortable	1	(2)	3	4	5	1	2	(3)	4	5

Fig. 2.12 Example questionnaire for gathering information data from customers relevant for the QFD matrix

The above questionnaires are collected and analyzed by entering the information on the QFD chart for our bike saddle example—in the left-hand shoulder, the Degree of Importance for the Customer and in the right-hand shoulder, the Customer's Comparisons Assessment with the competition (see again diagram in Fig. 2.10).

Establish Engineering Characteristics

The next step is to determine the Engineering Characteristics which must be optimized to assure customer satisfaction. The marketing domain tells us what to do, whereas the engineering domain tells us how to do. Engineering characteristics should describe the product in measurable terms and should directly affect customer perceptions. This translation of customer requirements into language that will be meaningful to a designer is a very important step in QFD: When it is not performed correctly, the major scope of the study—i.e., creating a product aligned with the customer's expectations—will be lost.

Look now at Table 2.7. It shows some of the major engineering issues determining the product characteristics that can affect, either positively or negatively, the perception of the bikers who will actually use the saddle. These ideas have been grouped under the three major headings, performance, cost, and safety, in line with the *VOC* subsection (i.e., the left side of table). For each customer requirement which emerged at either the secondary or tertiary sublevel in the data information grouping, the team has tried to make a causal link with at least one functional requirement by which the customers' perception can be influenced, and per each functional requirement identified, at least one, possibly measurable, engineering characteristic has been established to allow the engineers to proceed with the optimization process and assure customer satisfaction.

Once the engineering characteristics are established, they can be entered along the top of the QFD chart, while the metrics is entered along the bottom of the central relationship matrix, following the diagram in Fig. 2.10.

Engineering Characteristics Conflict Matrix

This stage involves filling in the relationship matrix in the main body of the chart and the triangular conflicts matrix at the top of the chart. The objective is to highlight the relationships between customer requirements and engineering characteristics, and conflict/supporting relationships among the engineering characteristics. Let us start from the relationship matrix.

For each customer requirement, namely for each row of the matrix, the team try to establish the level of causal relationship that can exist with each "Engineering Characteristic," namely the columns of the matrix. Or in product development language, the team did research into how strong the relationship is between the technical features of the product and the customers' specific requirements. The scoring of

Table 2.7 Main key engineering issues obtained by translation of major bikers' requirements for bike saddles

Voice of Customer			Technical translation of *VOC*		Metrics
Primary	Secondary	Tertiary	Functional requirements, FR	Engineering characteristics (Key Factors influencing FR)	
Performance	It should help to reduce loss of energy by pedaling	It should be light	Increase stiffness of the part	Maximum displacement produced by vertical force F	mm
		It should be rigid	Use thick sections	Cross section thickness	mm
			Use light materials , reduce volume	Weight mass	kg
	It should look "high performance" and aggressive	It should appear aerodynamic in shape	Use thin section	Cross section maximum thickness	mm
			Capability to vary profile sections (look for aerodynamic shapes)	Manufacturing process for very free shape design	–
	It should not cause pain after prolonged sitting	It should be comfortable	Dumping capacity/deform under cyclic load	Maximum width of material hysteresis under load cycle	mm/mm
		It should feel soft	Have a soft seat	Local pressure absorbed, p	MPa
			Respect the anatomical conformation	Central channel width, w	mm
			Reduce the pressure on the perineum	Height of back seat h_{sup}/depression height of the nose, h_{inf}/Central channel width w	mm
			Adjust for any seats	Length of the bar frame, l	mm
			Limit too much soft and thick padding	Local pressure absorbed, p	MPa

Table 2.7 (continued)

Voice of Customer			Technical translation of VOC		
Cost	It should have low operating costs	It should not require any maintenance by me	Achieve high durability	Salt spray corrosion test duration	Hours
		It should be durable		Prolonged duration test in saline environment without intermediate cleaning	Days
		It should be easy to clean	Can clean the materials using water	Wash test in water	None
			Avoid any nonconformities	Percentage of nonconformities	%
	It should have low purchase cost	It should be cheap	Cheap saddle	Material cost	$/piece
				Manufacturing costs	$/piece
				Management of nonconformity costs	$/piece
Safety	It is compliant with safety requirements	It should be safe	Respect the UNI EN 14766 on load cycles to failure	Cycles before failures	cycles
	It should be safe when it has a failure crash	If it crashes, I must be not injured	Avoid brittle failures	Deformation at failure break	Deformation at break under test

Primary	Secondary	Tertiary	Customer Degree of Importance	Maximum displacement produced by	Cross-section thickness	Weight mass	Prolonged duration test in saline environment	Cycles before failures	Maximum hysteresis under load cycle	Local pressure absorbed, p	Central channel width, w	Heigh of back seat h_{up}	Depression height of the nose, h_{nr}	Length of the bar frame, l	Salt spray corrosion test duration	Capability to clean materials by water
Performance	It reduces energy lost by pedaling	It is light	5	•	•				▽		▽					
		It is rigid	5	•	•	○		•	○		○			○		
	It looks high performance and aggressive	It appears aerodynamic in shape	4		•						○		○			
	It does not cause pain after prolonged sitting	It is comfortable	3							•	•	•	•			
		It feels soft	3							•						
		It is adjustable for any seats	4											•		
Cost	It has low operating costs	Avoid any maintenance by user	3					•							•	○
		Be durable	4					•							•	○
		Be easy to clean	3												▽	•
	It has low purchase cost	Be cheaper	3													
Safety	It is compliant with safety requirements	Respect the international specs	5					•								
		Be safe when it has a failure crash	5		○											

Fig. 2.13 Establishing a link relationship matrix extract from the bike saddle study

the strength level is conventionally made at a 0–9 discrete scale, in accordance with the original QFD matrix developed by Akao and Mizuno:

- 9, for a strong relationship (alternatively, we use the symbol "•");
- 3, for a medium relationship (alternatively, we use the symbol "○");
- 1, for a weak relationship (alternatively, we use the symbol "▽");
- 0, no relationship, no symbol.

As an example, in Fig. 2.13, there is an extract from the bike saddle study.

Conflicts Matrix

This is the triangular matrix at the top of the chart. It is used to highlight further supporting or conflicting relationships between two engineering characteristics. For example, if we look at engineering characteristics n.2 "cross section thickness" in Fig. 2.14, we note that it is identified on the top line by the symbol "▼", while some others are identified by symbols "▲" The former means that the engineering characteristic in this case must decrease if we are to move in the direction of improvement; the latter has the opposite meaning. This depends on the fact we want the cross sections of the saddle to be at the same time as thin as possible and to decrease in weight. However, the thinner they are, the more deformable the saddle, and this is also in contrast with customer requirement n.2 "the saddle should be rigid." The thickness of the cross sections is thus an engineering characteristic that must decrease to reduce the weight mass, but this can impact negatively on another engi-

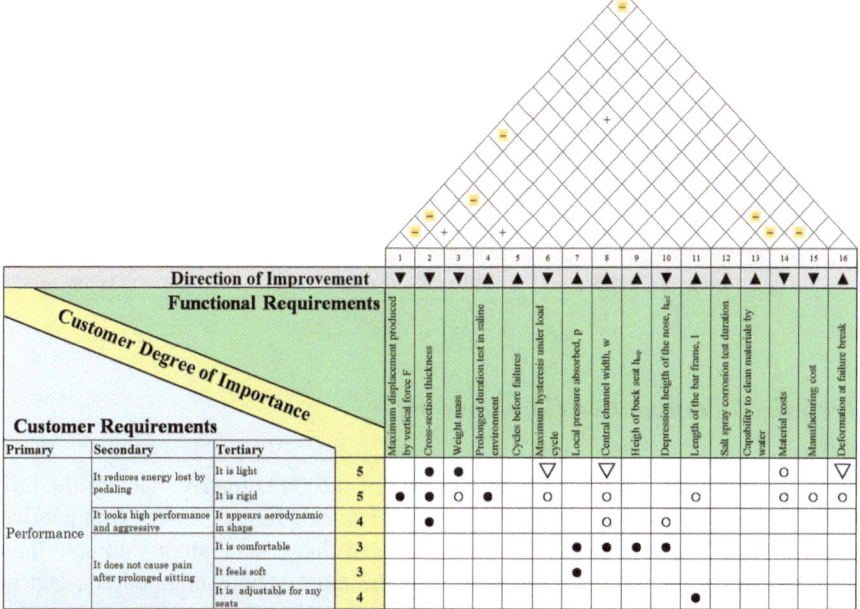

Fig. 2.14 Correlation matrix for the bike saddle study

neering characteristic that we identified—n.1 "Maximum displacement produced by vertical force F." Thus, we can state that when it decreases the n.2 engineering feature "cross section thickness," is:

- conflicting with the n.1 "Maximum displacement produced by vertical force F" that should be decreased and
- supporting of the n.3 "weight mass" that should be decreased.

All this information is symbolically summarized in the top triangular matrix. Symbols are used to indicate whether two engineering features are correlated by a positive supporting relationship or a negative conflicting relationship:

- for a supportive relationship, the symbol "+" and
- for a conflicting relationship, the symbol "−".

Negative symbols show where a trade-off situation exists, which must be resolved.

These will be considered in the next stage when the target values for the engineering characteristics are established. Below in Fig. 2.14 is an extract from the bike saddle study.

Table 2.8 Competitive comparisons on 1–5 scale and identification of suitable target values for n.2 and n.3 engineering characteristics for the bike saddle case study

Number	Engineering characteristics	Company	Measured data	Average value of measured data	Converted value on 1–5 scale	Target value metrics
2	Cross section thickness	A	4.2 mm	4.5 mm	3.2	3.75 mm
		B	5.5 mm		2.3	(score: 3.5)
		C	3.8 mm		3.5	
3	Weight mass	A	109 g	143 g	3.7	105 g
		B	140 g		3.1	(score: 3.8)
		C	180 g		2.2	

Technical Competitive Comparisons

The next step is to work on the bottom part of the QFD matrix, namely the box below "Metrics," which is indicated as "Technical Targets" and "Competitive Comparisons." Firstly, clarify the final scope of this crucial step: Compare how your product performs in comparison with the most serious competitors. But to make a technical comparison between your product and those of the competitors, you need:

- a basis of comparison to decide whether, as regards that specific engineering characteristic, your product is either better or worse than that of the competitors and
- such a basis of comparison to be as measurable as possible, to facilitate comparison.

The second issue has already been resolved since in the previous step *Establish Engineering Characteristics* you have pointed out the engineering features and their metrics. Thus, we focus on the basis of comparison, which is called in QFD language the technical targets.

Consider that the engineering characteristics are measurable; this is true for example in the cases of engineering characteristics n.2 and n.3 in the bike saddle case study, respectively, the cross section thickness and the weight mass.

Assume your engineering department has taken precise measurements for these two technical features on competitors' products A, B, and C. The results are summarized in column 4 of Table 2.8. In column 5, the average values for the data gathered are reported thus.

Now, we need a procedure to compare the competitors' products on a simplified semi-qualitative 1–5 scale and thus to identify at what level we want to position our product in relation to the specific engineering characteristics.

In Chap. 1, we already discussed two types of methods that help transform quantified values (by measurement or calculation) into a simplified scale, for example a 1–5 scale, in order to speed up the comparison process. We now focus just on the scoring-by-mean method and we apply it to this example. Thus, when we want to transform the data that have been gathered into a simplified scale, we need to proceed as follows:

- identify the average value of the gathered data;
- set that value correspondent to the mean value of the 1–5 scale, i.e., "3"; and
- transform the quantity measured into a 1–5 score value using the similitude above.

Before proceeding, it is important to highlight that when the values to convert are judged better as they increase, the relationship by similitude of average values is direct:

$$Converted\,Value_{direct} = \left(\frac{Value\,to\,convert}{Average\,Value} \right) \times 3 \qquad (2.14)$$

If we apply this relation to our data in Table 2.8, it would result in false scores, since the thicker cross sections would achieve a higher score on 1–5 scale. This depends on the fact the driving force is to minimize the thickness, and thus, the lower this value, the higher the score. We therefore have to use the inverted relation, which is:

$$Converted\,Value_{Inverse} = 3 - (Converted\,value_{direct} - 3) \qquad (2.15)$$

The above relationship simply allows us to invert the score in cases where we are dealing with lower quantified values to be highly scored. Equation 2.15 allows us to complete the data in column 6 of Table 2.8, reporting the final converted values.

Technical targets have been identified[14] in the last column, Column 7, of Table 2.8; setting the targeted scores on a 1–5 scale for each engineering characteristic (they are 3.5 and 3.8, respectively, for engineering features n.2 and n.3 in the table), we can inversely translate the simplified score we assign into original metrics values.

Now take some time to assimilate the above steps and pay attention to what is your main scope when you are constructing a QFD matrix for a product. So far, you have defined the "Technical Targets" in the bottom part of the QFD chart (refer to Fig. 2.10) for the engineering characteristics that the product you are developing should have in order to effectively answer the customers' needs and requirements. Now, we have the quantified values at our disposal, which we want to target in order to be successful on the market against our competitors. The last two steps in constructing the QFD chart therefore shall consist in:

- assessing our product and our competitors' product by comparing how close their respective engineering characteristics really are to such targeted values;

---

[14] In reality, choosing target values requires you to consider several things, such as the ratings and absolute values determined in the technical comparison with competitors, the importance ratings of the customer requirements you are trying to satisfy along with the associated data from the comparisons the customers have made with the competitors and any positive and negative relationships highlighted in the conflicts matrix.

Direction of Improvement			▼	▼	▼	▲	▲
Functional Requirements *Customer Degree of Importance*			Maximum displacement produced by vertical force F	Cross-section thickness	Weight mass	Prolonged duration test in saline environment	Cycles before failures
Customer Requirements							
Primary	Secondary	Tertiary					
Performance	It reduces energy lost by pedaling	It is light	5		●	●		
		It is rigid	5	●	●	○	●	
	It looks high performance and aggressive	It appears aerodynamic in shape	4		●			
	It does not cause pain after prolonged sitting	It is comfortable	3					
		It feels soft	3					
		It is adjustable for any seats	4					
Cost	It has low operating costs	Avoid any maintenance by user	3					●
		Be durable	4					●
		Be easy to clean	3					
	It has low purchase cost	Be cheaper	3					
Safety	It is compliant with safety requirements	Respect the international specs	5					●
		Be safe when it has a failure crash	5	○				
Metrics			[mm]	[mm]	[grams]	[hours]	[number of cycles]
Technical targets			40	3.5	105	200	2x10⁵
Technical Importance Rating			45	141	60	45	108
Relative Weight			5%	15%	6%	5%	11%
Our Product			3	2	4	2	4
Competitor #1			2	3	5	1	3
Competitor #2			3	3	4	5	3

Technical Competitive Assesment

+ Our Product

Fig. 2.15 Technical importance rating for a number of engineering characteristics for the bike saddle study

- rating our product and the competitors' using a method that can sum up in a proper way all the score values we obtained for the engineering characteristics assessed so far.

Let us proceed with first step and consider the extract from the QFD chart of the case study that is given in Fig. 2.15. The bottom part shows the row filled in with the "Technical Targets" and the rows (a piece of the chart diagram) relative to the "Technical Competitive Assessment" we carried out comparing each engineering characteristic of our product and those of our competitors with these technical values. The simplified 1–5 scale has been used and, lastly, a detailed assessment for each engineering characteristic has been performed for our product and those of our competitors. For the rating, we can proceed therefore to the next.

Technical Competitive Assessment: Rating the Engineering Characteristics

A combination of the customer importance rating and the strength of the relationships between the customers' requirements and the engineering characteristics is here established. As stated, the central relationship matrix (refer again to Fig. 2.15 that partly shows the relationship matrix or the simplified diagram in Fig. 2.10) gathers the correlation of symbols describing strength level, and numerical values are associated with these symbols: 9 for strong relationship, 3 for medium relationships, 1 for weak relationships, 0 in cases where there is no relationship.

The ratings are calculated by running down each column, summing the product of the customers' importance rating and the value assigned to the correlation symbol. As an example, consider the second engineering characteristic (second column of the engineering relationship matrix in Fig. 2.15). The absolute importance rating of 141 is calculated from: $(5 \times 9) + (5 \times 9) + (4 \times 9) + (5 \times 3) = 141$. The relative value in the cell just below is simply the absolute value expressed as a percentage of the total.

Again, if you consider the third column of the relationship matrix, the absolute value 60 is calculated from: $(5 \times 9) + (3 \times 5) = 60$, and its relative value, expressed as a percentage of the total, is reported in the correspondent cell below.

The last and most crucial step is the final scoring of our product and those of our competitors. We will follow the calculation procedure in the next part:

- for each product to compare and for each column of the relationship matrix, multiply each value in the technical competitive assessment grid by its correspondent relative weight value expressed in a percentage;
- sum all the results from the above multiplication.

Having carried out the above double-operations, we achieve for each product considered the results gathered in the "Total Weighted Score" box, on the right of Fig. 2.16. For example, for the "our product", the Total Weighted Score is calculated as: $(3 \times 5\%) + (2 \times 15\%) + (4 \times 6\%) + \ldots + (2 \times 7\%) + = 2.7$.

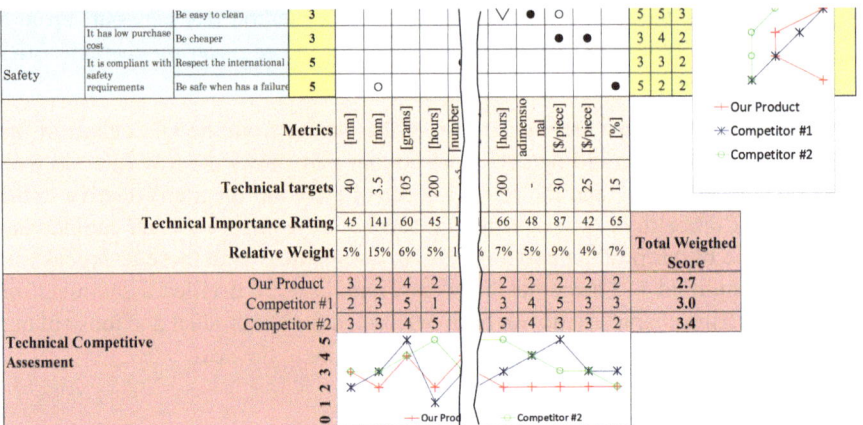

Fig. 2.16 Calculation of Total Weighted Score for three competing products in the case studied

The Analysis of Results

The data accumulated are finally organized in the QFD chart. Figure 2.17 shows the QFD chart of the bike saddle case study as an example.

An analysis of the chart is outside the scope of this chapter; in fact, the analysis of a traditional QFD chart applied to product development is outside the scope of the book itself. Some short notes are however useful since we will now be dealing with some general QFD analysis topics, when a special QFD matrix will be developed to support the material selection strategy.

Despite its rigorous method of construction, there are no set procedures for analyzing the chart. However, what even novices immediately recognize is that adopting QFD during the decision-making process gives them greater lucidity when in the final matrix they have to identify:

- The points that would improve, sometimes included in the product's specific engineering characteristics, but which are still missing despite all the competitors' efforts; the suggestion can sometimes rationally emerge that some expensive engineering features should be downgraded, because they are not relevant—or are not very cost-effective—in the customers' perception. It is an old story: Young researchers in the R&D department sometimes push for improvements, desiring to change the product and use materials that give higher performance and are lighter in weight, while their boss tells them: "Cool it! But nobody will notice. Nobody will pay out good money for this kind of change: they can't even see it!" Thus, what do you think about the brake calipers in sport cars, which, although they are of course functional objects, are more and more frequently left uncovered by the wheels, painted in bright colors and undergo research to give them captivating shapes?
- The most effective engineering characteristics, namely those that impact greatly on the customers' perceptions. Take another look at Fig. 2.15, and more pre-

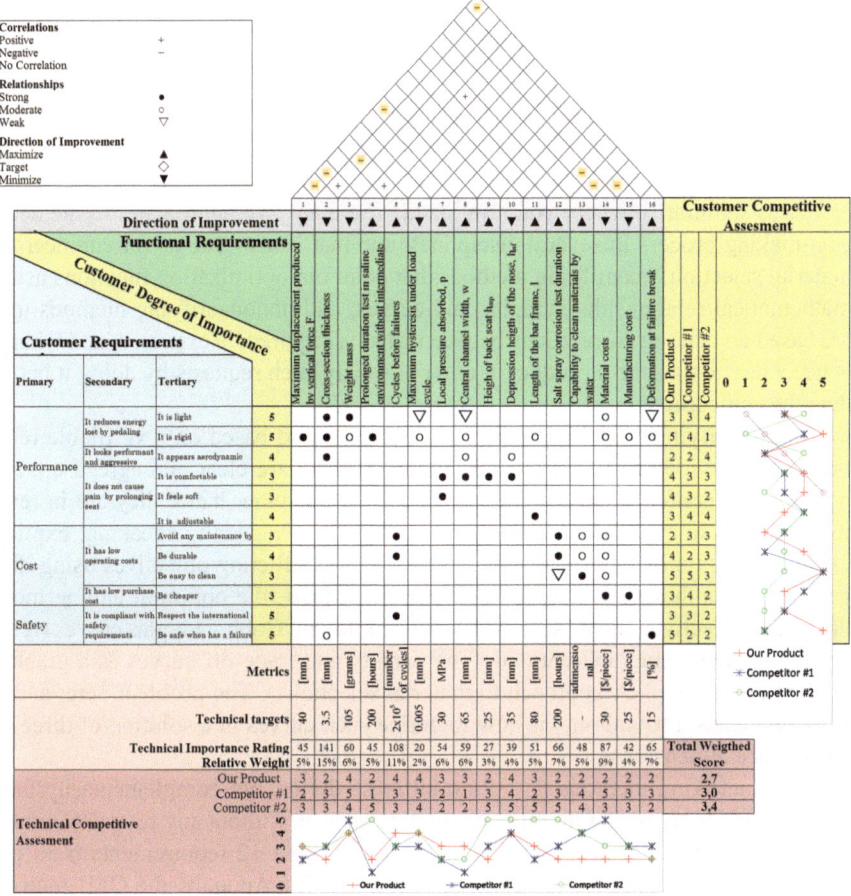

Fig. 2.17 Example QFD matrix for the bike saddle case study

cisely the row Technical Importance Rating, at the bottom of the matrix. And remember the values in each cell of this row represent a number that takes into account the strength of the causal relationship between the engineering features and customer satisfaction, weighted on the customers' relative importance. Thus, we found that out of 16 different engineering characteristics defined to answer customer needs, a pair of them, the second and the fifth, cover about 26 % of the customers' requirements.

- It can be easily used by a multidisciplinary team that can really work together confronting each other in free and open discussions.

Which Method—Explicit or Implicit—Is to be Preferred in Material Selection Strategy?

Finally, the ultimate question: what should I do? You still need to be a bit patient, just until the next chapter, before facing this question. And yet, so far, we really have paved the way.

Let us summarize where you now stand. There are two approaches to the decision-making process in several disciplines, and that is also true as for engineering material selection: quantifying methods that point out optimization functions using mathematical relationships, graphs, and curves; and nonquantifying methods that are based on making correlations between what is required, is expected, and what it is necessary to put into a product in order to satisfy such requests by doing it better than the competitors.

The quantifying methods are precise, effective, and based on a verifiable relationship, which, when the objectives and constraints are clear and agreed on, are not at the discretion of any single individual. On the other hand, they are in reality restrained to a limited decision-making "space." Practically speaking, explicit methods of optimization can manage a couple of conflicting objectives using diagram charts, although of course, theoretically at least, the optimum among more than three objectives can be searched for by employing Cobb–Douglas curves (refer to the previous paragraph entitled "The penalty and trade-off curves as a graphic method for the explicit approach to material selection"). The problem here, and it is by no means a trivial one, is how to define such curves in a solution of three or more space dimensions.

Nonquantifying methods do not do all these things, but complementarily they are suitable for analyzing multiple conflicting aspects without any restraint in their number: The bike saddle case study manages to target 12 requirements from the Voice of Customer and 16 engineering characteristics. Actually, the QFD matrix-based method can break the problem of optimization into small pieces, comparing and assessing solutions that have already been developed and then re-assembling them all into a final simple number, the Total Weighted Score.

In addition to that, we have to take into account the investment we need to make in flexibility, in terms of the high level of discretion left to users. QFD is just a tool, sometimes a software tool that is easy to use, but it does not do the job for you. While it forces the team to set target values for engineering characteristics in order to assure customer satisfaction, and it is able to show how you can impact with your choices and which leverages appear more advantageous, it certainly does not whisper in your ear and tell you how to do the job. In other words, it is flexible, but, if it is to be precise, it requires that all the experts that are competent in all the issues relevant to the customers' requirements and needs should join up and sit around the table it has laid out for them. QFD can act as a common language among all the diversities of a multidiscipline team that wants to take the product design as a whole. Using it puts people from each company department on equal footing, all able to stimulate the decision-making process, no one being sidetracked by partial views. Just like a round table, which, in a most effective metaphor, has no head, and at which everyone who has decided to sit does so with equal status and power for discussion. It is just democratic and perfectible. And since materials specialists sit at that table with all the others, why should they not bring some of the results they have obtained using implicit methods into discussion as well?

Chapter 3
The Total Explicit–Implicit Approach:
The Special QFD Matrix for Material Selection

Abstract **Chapter 3** faces the challenge of building up a tool that can embed the advantages of both implicit and explicit methods as discussed in Chapter 1 and Chapter 2, in order to mitigate their respective drawbacks. A classroom case study is employed in order to guide and support readers in preparing multi-objective decision analysis on the choice of material among possible contenders. Readers will therefore learn step by step how to customize the QFD tool for material selection which we call *QFD4Mat,* and how to embed, if necessary, tools that are usually of an explicit method like material performance indices. The method is provided in the form of a graphic tool that is user-friendly and easy-to-customize for the specific problem that the reader will face no matter what his own background; material specialists, designers, procurement officers, finance advisors and CEOs are all expected to contribute in providing their own ideas for the *QFD4Mat* to be created for specific projects. To facilitate the reader in customizing the *QFD4Mat* on his own, further electronic downloadable sources are provided in a website platform here presented. The basic matrix tool, the *QFD4Mat* employed in the case study discussed, other accessory sub-tools (e.g. material index lists, guidelines for defining the new product development process stage designation, customizable tables for calculating cost indexes for manufacturing and the Global Material Index etc.) are uploaded. The platform contains further classroom case studies realized by postgraduate master students.

In this chapter, we finally start moving from theory to practice. The next challenge we will be facing is therefore to build up a tool that can embed the advantages of the implicit and explicit methods as discussed in Chap. 1, thus mitigating their respective drawbacks[1]. However, before going on, I will risk becoming a bit boring by further repeating one basic principle on account of its importance in our next discussion. Just as the research carried out by the development team has to be based on insight about production needs in order to succeed in the market, so the process by which materials are selected will only be able to make good when and if it starts from the definition of a proper subset of material attributes that can positively impact product competitiveness.

---

[1] To this purpose, when necessary, we will frequently refer to basic foundations already laid down in the previous chapters.

© Springer International Publishing Switzerland 2015   69
F. D'Errico, *Material Selections by a Hybrid Multi-Criteria Approach,*
SpringerBriefs in Materials, DOI 10.1007/978-3-319-13030-9_3

As we learned in Chap. 2, the quality function deployment (QFD tool) can be a potent method for company managers and decision makers in that it provides a common language uniting all the diversities of a multidiscipline team that wants to take the product design as a complete whole. As stated, all the experts from the different departments can take part in the decision, because they can sit around the table with equal status and capacity of expression. Employing QFD in product development processes can help in seeking for a good compromise among the conflicting interests and competing objectives represented by each company department. Summing up, we learned that this is possible because the QFD tool provides an insight that is external to the company and therefore close to the customer: thus exploring product attributes—not solutions, remember!—that are vital for customers. One after another, as these attributes get defined, engineering characteristics can be identified that will translate the desired product attributes into technical features that are manageable during the engineering design process and can satisfy the needs and expectations expressed by customers in their own language.

We therefore need to define a framework, namely a suitable structure, by which we can organize the *key requirements* of the material so to create a positive impact on product functionality, usefulness, and value—three factors that will be essential to its competitiveness. A new framework is necessary because using a generalized list of material attributes similar to the one shown in Table 3.1 could not fit our purpose.

Organizing the Material Key Attributes

In the scheme of a QFD tool, *product requirements* are what we call the key attributes desired by a customer. Before we proceed, it is convenient at this point to clarify who our customer actually is.

For some products we may only have one or two types of customers, but there will more frequently be a *chain of customers*, where multiple customers take advantage of using our product.

For example, if we design and sell bike saddles, the chain of our customers would be as follows:

- the bike manufacturer
- the bike shops
- the online after-market sellers
- the user (the biker).

Remembering that one major scope of QFD is to capture precious suggestions from any customer we want to target, a complete approach should take into account the needs of each individual part of the whole customer chain. We therefore start to gather from our customer chain all the product *key requirements* that have any part to play in *product value creation* and that can be controlled by the choice of mate-

Table 3.1 General classes of material properties (Ashby 1999)

Class	Factor
Economic	Cost (price, for customer)
	Availability
	Recyclability
General physical	Density
Mechanical	Modulus
	Fracture toughness
	Yield and tensile strength
	Hardness
	Fatigue strength
	Creep strength
	Damping
Thermal	Thermal conductivity
	Specific heat
	Thermal expansion coefficient
Electrical and magnetic	Magnetic permeability
Environmental interaction	Oxidation
	Corrosion
	Wear
Production	Ease of manufacture
	Joining
Aesthetic	Color
	Texture
	Feel

rials. All such product *key requirements* can be organized into three categories as listed below:

- *Performance*, namely all those material attributes that can directly impact the functional aspects of the product. In market strategy language, these are attributes that are generally desired when pursuing competitiveness by technology-driven *differentiation*;
- *Cost*, namely all those economic aspects of a material that can directly impact on the trade-off between functionality and competitiveness on the marketplace. In market strategy language, these are attributes generally researched for pursuing competitiveness by *cost-leadership*;
- *Product attributes receptivity*, or simply *Receptivity*, namely the group or class of product attributes that have a subtle influence on customer product value perception.

Each of these three broad categories needs to be properly broken down into subsets. Table 3.2 shows an example of possible product requirements as driven by material attributes.

Table 3.2 Subsets of three categories for the material's outer attributes

Category	Category subset	Material outer attributes
Performance	Product life span[a]	The attitude of the material toward preventing the usual types of failure, that is toward counteracting damage modes (refer to the paragraph "The use of Root Cause Failure Analysis (RCFA) in material selection approach" later in this chapter)
	Functional objectives and constraints[b]	The material's ability to target functional objectives and constraints (e.g., light-weight design with stress, rigidity, thermal insulation, etc., refer to paragraph "Embedding Material Indexes" later in this chapter)
Cost and availability	Processing and fabrication factors[c]	The material's attribute in enhancing manufacturability (e.g., weldability, formability, hardenability, castability, etc.)
		Manufacturing process ability to target scale of production (e.g., die-casting for mass production with medium-/low-quality, gravity casting for lower production scale and higher quality, etc.)
Material costs		Raw material cost
		Restoring costs and sunk costs for nonconformities management
	Transformation process costs	Cost of secondary processes (e.g., bulk heat treatments)
		Machining costs
		Other supply-chain outsourcing costs (e.g., surface-hardening treatments, surface coatings, painting, etc.)
Receptiveness	Time for product availability	Delivery time
		Production cycle time
		Target production rate (e.g., is it a mass or customized production?)
	Reduced complexity/increased usability[d]	Features to be reduced below sector standard (e.g., oversized depths of hardened case used in manufacturing high load capacity linear guides subject to rolling contact fatigue can be reduced by gas nitriding case-hardening)

Table 3.2 (continued)

Category	Category subset	Material outer attributes
		Features to be eliminated (e.g., painting and following maintenance for corrosion attacks are avoided for bike tube frames when stainless steel is used instead of painted low-alloyed steels)
		Features to raise over sector's standard (e.g., *higher portability* has been pursued for laptops by increasing lightness and reducing section thickness because of the introduction of more expensive aluminum alloys instead of cheaper polymers)
		Features to be created (e.g., *bioresorbable* stents have been developed, thanks to bioresorbable and biocompatible magnesium alloys instead of conventional biocompatible metals, such as tantalum, titanium, or stainless steel, which remain in the body permanently until removed through further surgical intervention)
	Safety issues and compatibility	Material attributes consistent with the existing values, roles (e.g., low flammability and rapid self-extinguishing time (namely the capability of the material quickly to extinguish an ignited flame when the heat source is removed), two main requisites for civil aircraft cabin furniture)
	Ergonomics	Tactile feeling and sensitivity
	Observability and aesthetics	Texture and surface color variety

[a] Capability in withstanding load service conditions and mitigating damage and failure

[b] Capability in targeting functional (multi)objectives with (multiple) constraints

[c] The ability to form or shape a material falls under the processing and fabrication factors

[d] Degree of positive changes given by the *eliminate-reduce-raise-create* framework (Kim and Mauborgne 1997) of the *value innovation strategies* for the selection of key product features. Refer to Chap. 4

Why Does not the CEO Understand Me?

Company business units are organized into different hierarchical arrangements capable of varying degrees of coordination or integration with the enterprise's key functions. A number of these share some key functions with other business units or rely on centralized corporate services, such as R&D, for functional support. We have neither time nor space here to take an in-depth look at strategic business organization and leave that to the specialized literature and master courses, only borrowing a few basic principles for our own purposes (which are based, furthermore, on common sense and need no specialist background in management).

An organization is composed of structured units, each of which is a center of responsibility for producing an output by transforming some inputs[2] in order to take part in the business strategy of the organization. For the success of the organization, it is vital that such plans coming from the different business units are all well inter-coordinated. For example, the production area needs to coordinate with the sales area to ensure that the production processes achieve the expected volume of sales. Similarly, the raising of short-term funding must be based on forecasts of cash flows—inflows from sales as well as outflows from operating expenses. However, in reality it often happens that some units' short-term objectives—1 year more or less—are not fully aligned and coherent with the long-term organization strategy. It should be noticed that organizations themselves do not actually have goals; it is the people that make these up in order to have their own goals. And very often the people's objectives are different and not always perfectly aligned with the business strategy of the company. For example, if the system excessively emphasizes the need to reduce costs without adequately specifying the constraints, and a manager responds to this input by effectively reducing costs, but at the expense of quality, then it could be said that the system seems to have motivated the manager in the wrong way.

To conceptualize, we can say that when a cost-based strategy is pursued, relationships among R&D, marketing, sales, and production functions are normally fairly straightforward. It seems so simple: production is asked to manufacture at a low cost and the sales people sell and compete by basing themselves on the low price. In fact, in such cases a simple cost competitive product is the minimum necessary to participate in "the game." But quite apart from the simplicity or complexity of the product offered, as far as the potential buyer is concerned, a product is a complex cluster of value satisfactions. And exploiting this complexity is the essence of a differentiation strategy.

Now put on the hat of your company's CEO.

You as a manager contend that differentiation generally involves a major challenge against less predictable factors such as innovation, customer preferences,

---

[2] There are four types of responsibility centers or business units: the revenue centers where the outputs are measured in monetary terms, the cost centers where inputs are valued in monetary terms, the profit centers where both inputs and outputs are measured in terms of money, and investment centers where performance activities are assessed in relation to one or more investment activities.

cooperation from distributors, etc. This unpredictability means you have to face uncertainty. As uncertainty increases, the manager usually demands more information; or rather, and more precisely, he demands more quality information from the business units. This generally works up anxiety in the business units' managers, resulting in detailed internal reports being drawn up, which describe only one piece of the story. Contradictory organizational arrangements, interests, budget constraints, and assigned targets are implicitly contained in the report, despite the fact that there is a general strategy to pursue.

For example, contradictory arrangements and decisions would be pursued simultaneously on account of the fact that the sales business units are close to their customers and externally focused on the opportunities to differentiate the product using a new type of material, while, at the same time, the production manager and buyer are internally oriented toward a hardline product cost reduction to be pursued by searching for cheaper raw materials, outsourcing heat treatments, and discouraging the search for new materials because of the investments necessary to make in order to adjust the production line. To present the very worst scenario you could ever have, assume that the R&D team is working in their ivory tower taking all their technical decisions on materials, their treatment, and processing in a quite isolated way. They are pursuing the objectives of creating the best performance and are personally being driven by straight benchmarking among the technical features that the competitors offer. As a result, a confusing strategy is developed, the result of conflicting interests and objectives that various people pursue inside their own organization unit. And it also very frequently happens that some key decisions for product development are simply taken at the coffee machine by a very few individuals. The above situation could lead to a business disaster in three moves:

1. the R&D pursues for the best performance product in order to hit the competitors by introducing some technical refinements;
2. such technical refinements are put into products because new investments in production have been made that raise costs;
3. the final price rises put the sales manager is in a quandary since customers might be not ready to pay an extra price for these new products.

To complete the negative scenario, only after production (i.e., investment) has started, does the marketing manager realize, when he gets to see the first sample produced, just how little care and attention have been given to the information on customer needs and segmentation provided by the marketing area. The engineers judged that some technical features were vital in order to distinguish the product on a measureable basis, but unfortunately these would turn out to be insensitive to users[3]. In some lucky cases, the business units' major objectives are aligned with

---

[3] It is worth to further notice that the cost of new product development increases for each day lost because of poor interdepartmental cooperation. For this reason cross-business-unit collaboration is central to a large firms' value creation. Emergent theories emphasize that a *BU-centric process* led by "multibusiness teams" of general managers leads to a better collaboration than a corporate-centric process.

Fig. 3.1 Example of crankshaft component for automotive application

company business strategy. This could be the case, for example, in a manufacturing company's strategy that orients toward product cost reduction and that is pursued by the buyer looking for cheaper materials to acquire. In fact, a buyer will not generally have any specialized background in materials, and delegating critical decisions regarding the procurement of raw materials could be uneconomic in the long run. Let us consider an example.

A manufacturing company is preparing its best commercial offer for a big customer, for example an European carmaker, leader in the automobile sector, to develop, manufacture, and deliver a new series of motor engine crankshaft components (Fig. 3.1). The product manufactured so far was made in carbon steel, with a 0.45 % carbon content, acquired in the form of round bars. The unitary cost for the material is 0.8 €/kg. In the conventional production process, the round bars are subjected to several hot forging steps to shape the material into a semifinished profile. The forged shafts are thereafter machined, but further tolerances are left for the following critical step. The semifinished shaft moves therefore, from the machining stations toward a heat treatment unit—inside the company—where the semifinished part is quenched and tempered to increase its mechanical properties. During quenching, some components experience slight distortions induced by the quenching step, namely nonconformities that are actually fully recoverable by cold press straightening and new machining operations. Because the component is required to resist loads, some portions of the shaft surface need to be hardened. For this purpose, a localized surface hardening step is therefore applied to increase the surface hardness. This process is very critical since some surface cracks can occur during the surface water quenching. On a yearly production basis, they always get 1 % of defective components that are not reparable. These components have to be discarded when such cracks emerge during the quality check. If not detected, the cracks would be responsible of premature failure during the product's service life.

The last steps of a conventional product cycle are machining and finishing operations, final packaging, and delivery to the customer (the carmaker) for assembling.

During the closing technical meeting, an alternative in material and processing to reduce the 1 % product nonconformities is explained by the material specialist to the manager. The proposal consists in changes to the raw material and to the surface

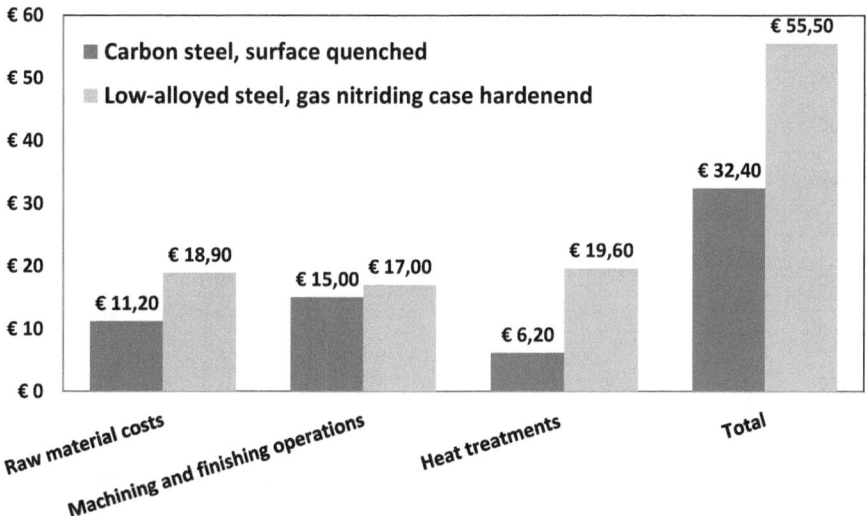

Fig. 3.2 Product cost structure for a component in two case scenarios: surface quenched carbon steel and surface nitride low-alloyed steel

hardening process used. The material specialist illustrates that a low-alloyed steel, of 0.40 % carbon, around 1 % chromium, and some molybdenum, could give a positive result in reducing the surface crack risk, if the pieces were properly dealt with using an alternative surface hardening treatment, namely a gas nitriding thermochemical process. This treatment would be conducted at low temperatures and would achieve the surface hardness targeted without any collateral problems: neither distortions nor surface cracks would occur. The treatment should be outsourced.

The proposal sounds really interesting to the manager, who asks for some numbers regarding the changes to the cycle. Numbers, in the manager's language, mean the unitary cost of product to be sold. As a reply to the question about raw material unitary cost, the material specialist comments that a low-alloyed steel is around 1.35 €/kg; this means an increase of around 68 % over the carbon steel currently used (this first reply would provoke some seconds of silence from manager). The second question is how much is the outsourcing cost of the nitriding process. As the outsourced nitriding process starts from 1.1 €/kg—excluding packing and delivering—the unitary cost accounts 19.6 €/piece, against more or less 6.20 €/piece for the internal surface quenching (such a second reply could provoke final oblivion for the material specialist). Figure 3.2 summarizes the differential costs in the two scenarios.

As against this, the material specialist could have told a very different story if some data on the costs of product nonconformity management had been gathered with the help of the production manager and the sales area manager. For example, the material specialist could illustrate that conventional carbon steel can be subject to surface cracks induced by water quenching that can occur to the fillet radii of the component and to other critical geometrical points. Such nonconformities gen-

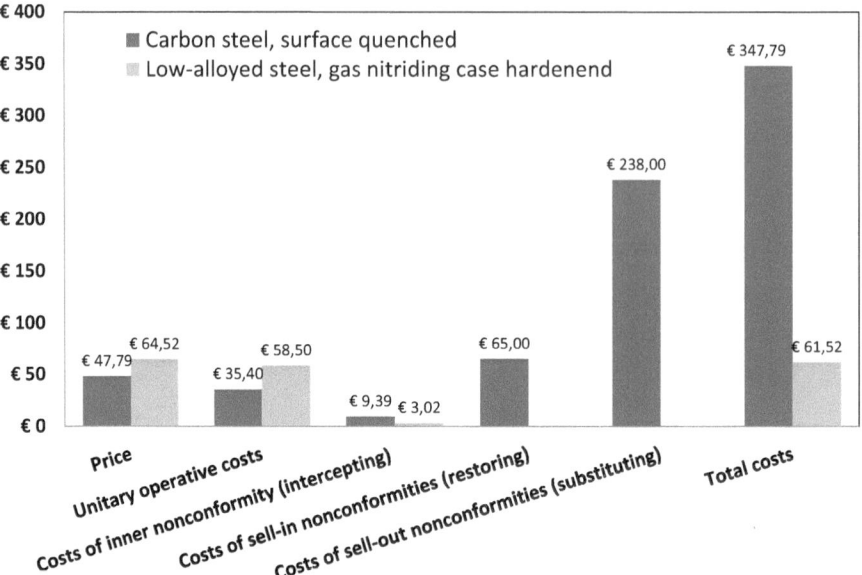

Fig. 3.3 Comparison of unitary price versus operative costs. Costs for restoring product because of nonconformities are accounted in three phases of the supply chain for surface quenched carbon steel and surface nitride low-alloyed steel products

erate further costs for each product that has to be overhauled, costs that increase substantially if the defects are intercepted during the assembling phase, or in the worst case—when not intercepted at quality check gates—if they remain invisible and embedded until a long-term failure of a component occurs during product service. The material specialists can illustrate this scenario and the disastrous impact it would have on paying the costs of damages, which would be much higher than the selling price. As emerges from Fig. 3.3, the total costs for nonconformity management vary substantially depending on the stage of the supply chain and delivery at which the surface cracks are detected and the component reworked.

The material specialist would claim that an outsourced nitriding process is more expensive, but it involves less critical final treatment for the component. To really evaluate the cost and benefits of a change in material, it would, therefore, be necessary to analyze the impact on direct variable costs, on the one hand, and the positive reduction in costs that would derive from delivering a more reliable and higher quality product to the customer[4], on the other hand. Thus, completing the

---

[4] This approach arises from the more general life cycle cost accounting (LCCA) approach. Life cycle cost (LCC) analysis initially applied by the US Department of Defense (Gupta 1983) today covers wide aspects of product life cycle engineering. It aims to seek out the "hot spots" of the whole product life cycle in terms of saving costs in order to calculate the true cost of the material components of products, their use, and ultimate disposal or recycling throughout their

snapshot, an overall unitary nonconformity cost would be foreseen for two alternative scenarios as depicted in Fig. 3.3. A total evaluation of gross margin over total supply should, therefore, also take into account the direct economic impact for product substitution and the indirect economic impact on product satisfaction in the customer (performance, safety, and out-of-service costs result in a considerable financial loss to the system user).

What therefore emerges from this classroom workout is something that happens very frequently in real cases. It is only common sense: the selection of materials and processes for the manufacture of a product tends to focus primarily on the unitary cost of the materials. Products need to be sold, thus pricing strategy is vital for a company to survive in a competing global marketplace. As marketing area managers, we know the story. But if we put on the hat of an R&D engineer, it is difficult to accept this need to manufacture products using a combination of materials and processes that may lead to a substandard quality and reduced operating life. In the long run, this could lead to financial losses as a result of having to iron out nonconformities to litigations, management and product liability in lawsuits, and to a reduced brand loyalty for the product in more serious cases.

A major and difficult challenge that we are setting up in this book is to comprehend that all points of view, those of the development engineers, the commercial agents, the marketing experts, all need to be considered in depth. The litmus test we can use to check whether we have effectively contributed to an optimal choice for our company to take is when we finally stand up at the table and feel completely satisfied, despite we achieved a mediated position we did not expect to reach when we began taking part in the discussion (when we were convinced to remain inflexible to any negotiations!)

The Use of Root Cause Failure Analysis (RCFA) in Material Selection Approach

As emerges from the previous case study, a concurrent advantage in the selection of more expensive materials sometimes consists in the mitigation this will bring to product nonconformities, especially with regard to major nonconformities (e.g., anticipated product failures, reduced usability of the product, etc.) that our administration might have to manage in future. Sometimes critical nonconformities can result in hazardous or unsafe conditions for individuals; thus, in the long run, we

---

lifetime. Life cost cycle accounting serves therefore to actually compare similar products or processes to determine which is the more efficient. It targets the breakdown of all the costs incurred in the product's life from production and construction costs (manufacturing, process development, production operations, quality control, etc.), the initial logistic support requirements (e.g., initial consumer support, the manufacture of spare parts, the production of test and support equipment, etc.), the operation and support costs (e.g., marketing and sales, transportation, corrective maintenance), retirement, and disposal costs.

have to consider the very serious consequences that might result, the most dramatic of which could even bring the company to a downfall.

The Japanese concept of *poka-yoke* (literally "mistake-proofing") was developed in the early '60s, and its basic principle is the basis of all methods that have been further developed and are now in use to seek and identify problems as close to the early design and test stage as possible, in order to eliminate product defects. The best results come when a new product is designed to prevent and eliminate sources of premature failures. In the engineering field, the *Failure Mode and Effects Analysis* (FMEA) is one method that examines potential failures in products or processes (today FMEA is also employed by service providers, as for example, for planning the delivery of health care services by private or public companies). It helps select counteractive tasks that reduce the impacts of life cycle consequences of a system failure. Not going too deep into the matter here, what is important to highlight is that an FMEA consists in a breakdown of the components of a system into functions, analyzing possible *failure modes* and their related *causes* in order to estimate their impacts (i.e., *failure effects*) and to put preventive measures into action.

What engineers call FMEA when they study preventive measures for avoiding premature failures in systems, products and processes, is the same thing that material specialists call *Root Cause Failure Analysis* (RCFA), during which similar analysis is applied to materials. In order to grasp the main concept, let us return to the scene where we, as customers, are taking a look in a bike shop before purchasing a new bike. We are focusing on its attractiveness and, if we are particularly keen, on its technical features, as reported in the product info panels. But let us try to change our point of view and consider ourselves as the designer of the frame instead—namely the one who took decisions about design features in order to try and make sure that the bike would effectively communicate to the customer: "Buy me! Buy me!"

As designers of a medium–high-level bike, we accepted that bike enthusiasts would be taking into account the efforts we made in weight saving, regarding both shape creation and choice in using lightweight materials. And we accept as bike designers that it is not our job to deal with the fact that all our efforts in giving bikers what they need—reducing our product by a gram or two under those of our competitors—could easily be nullified if we as cyclists frequently stop at a hot dog cart near Central Park after our rides in the sun.

What we have to face are the problems of choosing a material from among old-fashioned but still so reliable steels, light aluminum alloys used in aeronautics, titanium, the metal of the gods, magnesium as light as plastics when we handle it, or the lightest carbon fiber reinforced polymer materials used in top cars. Before launching on any matrix scoring, we have to focus primarily on a single question: what wall thickness should we chose for the tube? Actually, here, we are really trying to answer the fundamental question: how strong must the frame be in order to work safely during the bike's lifetime?

Before moving toward calculations and computer-assisted designs (as explained in the last paragraph of Chap. 2), we can turn the problem round by backtracking a moment and starting from what we know—or we can easily find out from statistics—about how likely it is for a bike frame to fail.

Paradoxically, it is not so usual to start analysis from the identification of the root causes of failure in components, namely from the end of the story. Yet, it comes so naturally and intuitively to any nonexpert to imagine what would happen in a case where—by obeying only our bikers' weight saving requirements—we have chosen a tube frame made of a thin foil of aluminum. I guess nobody would ever agree even to get on such a bike, instinctively held back by the perception of the risk involved that the tube might collapse as soon as it takes the load. What is intuitively introduced here is that one main *key feature* guiding *material selection* for the tubes is its capacity to resist an external load without braking. Thus, the *ultimate tensile strength* (UTS) is the key feature that measures the load per unit area of a tube section; another key feature of a material will be the one that prevents irreversible deformations incurring in the tube, in which case it would fail in its main function; the yield strength (YS) is the key feature to estimate acceptable load per unit area of tube section that it can undergo without plastically collapsing. These two key features, the UTS and YS, are strictly related therefore to the resistance properties of the component when a static—that is a noncycling—load is applied. By taking them into consideration, the designer avoids unexpected failures to the tube function due to an underestimation of material overloading.

On the other hand, as bikers we do not like the tube frame to bend excessively or elastically and to twist under the load we transfer by our pedaling; it gives us a bad feeling when we realize that a significant part of the energy we spend in pedaling is wasted by the cyclic, reversible, namely elastic, deflection of the tube frame. The *key feature* that measures the resistance of the material to elastic deflection or bending is the *elastic modulus* or *Young Modulus*. Steel is chosen because its elastic modulus is high; aluminum has an elastic modulus that is 60% lower than steel, but it is also 65% lighter than steel. This allows the designer to explore larger tube sections (i.e., acting on thickness and/or outer diameters) to increase the *rigidity* of the tube itself and balance up the additional material used due to aluminum's lower density. For this reason, aluminum tubes in bike frames are bigger than steel tubes, because similar rigidity is thus achieved, and, in some well-designed shapes, the aluminum tube frame is more rigid and lighter than a steel frame. High yield strength, high tensile strength, and high modulus (Fig. 3.4) are three selection criteria for our tube frame, but that is only part of the question.

On the assumption that the tube is lighter and thick enough to resist buckling, it is right to say that the idea of resistance focuses on the fact that a bike frame needs to survive all along its whole service life. It shall therefore support several *load cycles* induced by both pedaling and ground reactions. The bike frame is expected to resist the *time-dependent damaging* processes to which materials are subjected when they are exercised for several repeating cycles. Engineers call such failures *fatigue failures*. They occur in a certain modality that can vary slightly for each class of materials, but they generally consist in the appearance of an initial surface crack after a certain lifetime period. Not being visible to inspection by the naked eye, such a crack can dangerously propagate under loading cycles without any external indication, progressively reducing the tube cross-section area, which is solicited by the internal stresses induced by the external load applied. Once the area that has been reduced by the crack propagation can no longer take the following load cycle,

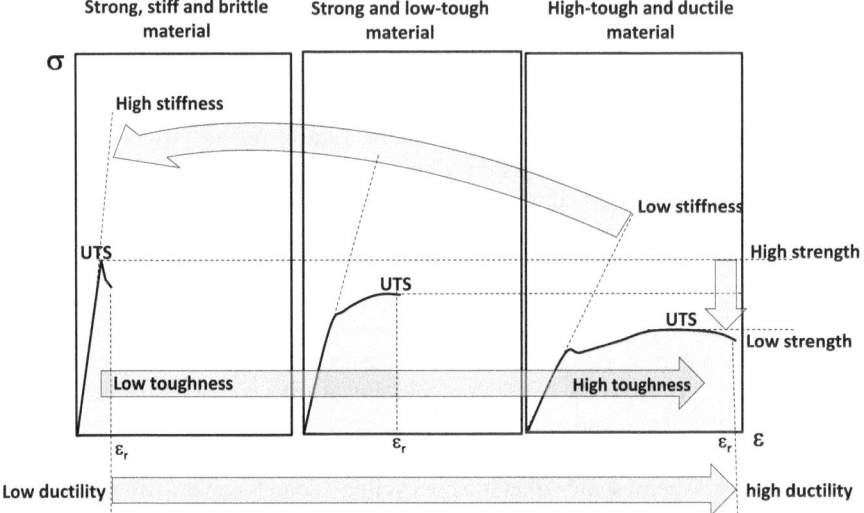

Fig. 3.4 Understanding basic material properties and their mutual correlations. For example, a high stiffness material (scheme on the *left* side) is a material with high slope—measured in the origin—of the σ (stress) and ε (deformation) curve obtained by a tensile test on the material. This slope is called the elastic modulus of the material. Such a high stiffness material usually exhibits further high mechanical strength (i.e., ultimate tensile strength, UTS), low ductility (i.e., deformation at break, ε_r), and low toughness (i.e., energy stored to break, proportional to area below the σ–ε tensile test diagram)

the tube breaks. It is worth noticing that the tube breaks without any forewarning on account of the absence of any outer signals, for example deformation, during the gradual advancement of the crack.

Is that all? Well, no, it is not. To tell the true story, if we are planning an amateur session of downhill mountain biking, there is at least one other important issue to consider. Downhill bikes must be heavy, strong, and feature front and rear suspension with over 8 in. of travel, to glide quickly over rocks and tree roots. But what should be the best material for the front forks?

Some years ago I was involved in an investigation of a serious accident that occurred to a downhill biker. The suspension forks incurred in a break that was as fast as it was unexpected during the ride. The brace (or arch), namely the linking bridge located between the two lower legs of the forks, catastrophically ruptured. Examining the surface fracture, it was clear the rupture was without any plastic deformation, typical of a *brittle failure*, abnormal for such a safety component. Unusually, the material chosen was too brittle for the application, as a deeper analysis revealed. Large interconnected porosities, actually big voids, produced during a low-quality casting process had localized near the arch of the suspension forks system. A big mistake was made by processing an intrinsic brittle material, a high-aluminum content magnesium alloy, with the wrong casting parameters. That *failure* provoked serious injuries to the biker, fortunately almost solved after several months, but it could have been really catastrophic. Brittle—or embrittled—materials can have high yield strengths, and an apparently ordinary tensile strength, but they have very little plastic deformation capability. If defects are present (e.g., porosity embedded during the manufacturing process, the imperceptible cracks formed on the surface, etc.), the reduced plasticity at the crack tips and locally induced stress can reach such high values that they are literally able to quite instantaneously break apart the interatomic bonds in the material. The fracture crosses the component in milliseconds, so that this failure mode is called *fast fracture* damage. This is what happened to the biker's suspension forks: the material used had a very low *fracture toughness* or K_{Ic}, namely the *key feature* that measures the capacity of a material to resist to stress below yield strength in the presence of a defect without incurring in *fast fracturing* failure. I have a vivid memory of that case study, so a forged and not a cast product, I guessed, could be a good choice for front suspension forks.

Lastly, we might do a brief evaluation on environment conditions, as common sense will tell us that *corrosion* damage can sometimes develop at critical points, especially in cases where frames are used offtrack and undergo scarce periodic cleaning. There are several forms of corrosion damage, but more frequently bike components develop uniform corrosion on the surface. This can provoke aesthetic problems that are usually overcome by using protective layers or coatings. The problem of *uniform corrosion* is generally treated by measuring the loss of material in particular environmental conditions, that is, the penetration of corrosion[5]. A user-friendly key feature to control the effect of penetration is the *rate of corrosion*[6], namely the speed at which a metal deteriorates in a specific environment. The rate, or speed, is dependent upon environmental conditions as well as the type and condition of the metal.

A structured RCFA approach to identify such material key features, essential if we are to avoid or mitigate premature failure and extend lifetime to expectation, generally consists in three main phases, as summarized in Table 3.3. The purpose of *Phase I* is to identify the main damage mechanisms that would incur to the material by carrying out an *ex-post* analysis, for example studying damaged components or gathering statistic data from failure case databases, or by *ex-ante* analysis, for example analyzing the environment and the foreseen load condition. As shown, seven

---

[5] For a bike tube frame cases of localized corrosion mechanisms that appear to the naked eye are not so frequent as microscopic surface voids. This form of corrosion is called *pit corrosion* and consists in highly localized loss of material that progresses over time because of an aggressive environment. If the tube—or component—subject to pit corrosion is loaded under cyclic tensile stress, for example, in tubes that contain pressurized fluid, such microvoids could also become preferential sites for the nucleation and propagation of cracks. The result is a change in the "evolution" of damage mechanisms, which progress as *fatigue cracks* induced by *pit corrosion*.

[6] Corrosion rates in the USA are normally calculated using mpy (mils per year equivalent to 0.0254 mm/y = 25.4 μm/y in International System of Units). In order to calculate the rate of corrosion, the following information must be collected: (a) weight loss, that is the decrease in metal weight during the reference time period), (b) density of the metal; (c) total initial surface area of the metal piece; (d) the length of the reference time period. A very user-friendly, free corrosion calculator and multienvironment database are present at http://www.corrosionsource.com/FreeContent/Tools-Home.

Table 3.3 General phases of material key features to prevent or mitigate failures by an RCFA approach

Phase	Task
Phase 1—Gathering information on failure mechanisms and damaging phenomena involved	Gathering all the necessary system information, such as:
	Physical parameters and constraints;
	System geometry;
	System environment;
	Damaging processes. The presence of one (or sometimes more than one) of the following five classes of macroscopic damaging phenomena can be addressed:
	Plastic deformation and break by overload due to overpassing the yield strength or tensile strength;
	Corrosion damage due to chemical interactions with the environment
	Wear damage due to friction between two surfaces
	Fatigue damage caused by cyclic loading
	Surface fatigue damage caused by cyclic loading acting on top surface
	Fast fracture damage due to impact loading that can cause materials to brittle fracture under specific circumstances
	Creep rupture damage due to the accumulation (time-dependent failure) of plastic elongation over time, when metals are subjected to low stress applied at high temperature (above 0.5 of metal melting temperature)
Phase 2—Pointing out key factors for failure	A *cause–effect relationship* among properties (i.e., material *key features*) and failures is established in order to select such key features that can counteract **key factors** governing damage (refer to Phase 3).
	(Example: in the vehicle crankshaft study case introduced in Chap. 1, see Table 1.9, once failure fatigue and wear have been identified as two main damage modes that affect such a component, it is possible to pinpoint *surface hardness* and *surface finishing* as two main key factors that can mitigate both failure modes)

Table 3.3 (continued)

Phase	Task
Phase 3—Ranking possibilities	A *relationship matrix*—similar to the one shown in Table 1.9, Chap. 1—is set up, scoring the relative weights for the key features of materials that can mitigate relevant failure modes
	(Example: it is known that wear damage—at constant lubrication parameters and load—can be mitigated by increasing hardness on top surface. And that the depth of a hardened surface layer is the key to controlling damage mechanisms that are
	typical of repeated stress on the top surface layer developed by cyclic loads exchanged by cyclically counteracting bodies that come into direct or indirect contact through an interlayer of lubricant film,
	that is *fatigue contact* mode. Alternative materials and treatment options are therefore scored on the basis of their ability
	to mitigate and counteract failure modes occurring to the product. Having been multiplied by the weights assigned for each failure mode,
	they are lastly ranked and included in the final score matrix:
	refer back to the first three rows of Table 1.9, which address the wear resistance, the fatigue resistance, and resistance to contact fatigue of the three options)

main damage modes are recognizable for materials: (1) plastic deformation and break by overload, (2) corrosion damage, (3) wear damage, (4) fatigue damage, (5) surface fatigue, (6) fast fracture damage, and (7) creep rupture damage.

The scope of Phase II of RCFA (Table 3.3) is to identify which material key features you need to control in order to avoid or at least mitigate failure modes. The output of Phase II consists therefore in realizing a correlation matrix among mitigating actions of damage modes and the material key features to be controlled to that scope.

The *seven failure modes* are causally correlated with *material key features* for failure mitigation as shown in Table 3.4. The material key features are properties (we call them indifferently key features or key attributes) of a material that are measureable by tests and that primarily take part in controlling failure mechanisms. The symbols (+) or (−) next to these key features indicate respectively when an increase (+) or a decrease (−) in the value of the specific key feature is beneficial to damage control. In some cases some primary key features can also be correlated with secondary key features, namely those material properties that have a further influence on the failure mechanism. For example, fast fracture damages are sensitive to parameter K_{IC}, the measured fracture toughness of a material; thus, the higher the K_{IC} value, the lower the risk of fast fracture. At the same time, the lower the yield strength, generally the higher the ductility in metals; high ductility material can absorb greater deformation energy at the crack tip of the defect, thus resulting in an increase in the total fracture toughness of the material.

It is beyond the scope of this book to go into depth on material *failure modes* and the *damage mechanisms* that govern such failures. The intention here is dual: to provide a general scheme for the material specialist that works in your multidisciplinary team; to communicate the principle that it is not so difficult for the material specialist to identify potential critical failures that could occur in the presence of foreseeable loads and work conditions on the one hand, and, on other, the specific material key features that can counteract the damage mechanisms involved.

The correlation matrix between failure mode and material key attributes will be useful when we want to construct the customized QFD matrix to select the optimal material for our product, i.e., when we want to correlate product requirements related to the service, lifetime, and the key attributes of the material to be assessed. Returning to the previous example, the crankshaft shown in Fig. 3.2, a possible part of QFD that regards the product life span extension, can be constructed as shown in Fig. 3.5.

Note this: compare Fig. 3.5 with the previous Fig. 2.15. The two portions of QFD matrices use the same framework. Actually what distinguishes the QFD for material selection, which for simplicity and quick reference we will be calling *QFD4Mat* in future, is how the product requirements are identified and thus ordered. Particularly, exploiting the general scheme shown before in Table 3.2:

- the first column of *Product Requirements* identifies the *Performance* features of the product;
- the following columns report the subset of *Performance* features that would impact *product functionality*, particularly product *life span* and *functional objectives and constraints*.

Table 3.4 Correlation among failure modes and key features of materials that have mitigating effects on damage mechanisms

Failure damage	Key features	
	Primary	Secondary
Plastic deformation/break by overload	(+) UTS [MPa], tensile strength[a]	(+) KV or KCU [Joule], notch toughness[c]
	(+) YS [MPa], yield strength[b]	(+) E%, percentage elongation (also defined deformation at break ε_r, see Fig. 3.4)[d]
Corrosion	(−) K [mpy], corrosion rate	—
Wear	(+) Hardness (e.g., HV in Vickers test, HRC in Rockwell test, etc.)	—
Fatigue	(+) S_e [MPa], fatigue strength[e]	(+) YS [MPa]
		(−) Ra, surface roughness
		(+) Effective case depth (in case of surface hardened component)
Surface fatigue	(+) H, Hardness (on surface)	(−) Ra, surface roughness
	(+) Effective case depth[f]	
Fast fracture	(+) K_{Ic} [MPa·1m]	(−) YS [MPa]
Creep rupture	(+) Tm, melting temperature [°C]	—
	(+) Q, activation energy for creep [KJ/mol]	

[a] Static resistance limit
[b] Elastic limit
[c] Amount of energy absorbed by a material during fracture: it is a measure of the ductility property of a material
[d] Percentage increase in length of material sample after fracture in tension test—it is a measure of the ductility property of a material
[e] Also called endurance limit. For steel, there is a lower stress limit below which fatigue samples that have been surface polished and with no defects do not fail. Usually this stress limit occurs in steels by around 10^6 cycles; for other metal alloys, such as aluminum, that do not exhibit this behavior, conventional fatigue tests are conducted to seek the stress level for achieving at least 10^7 cycles
[f] The effective case depth is the depth of the case that has a hardness equivalent of HRC 50 (equivalent to around HV 510)

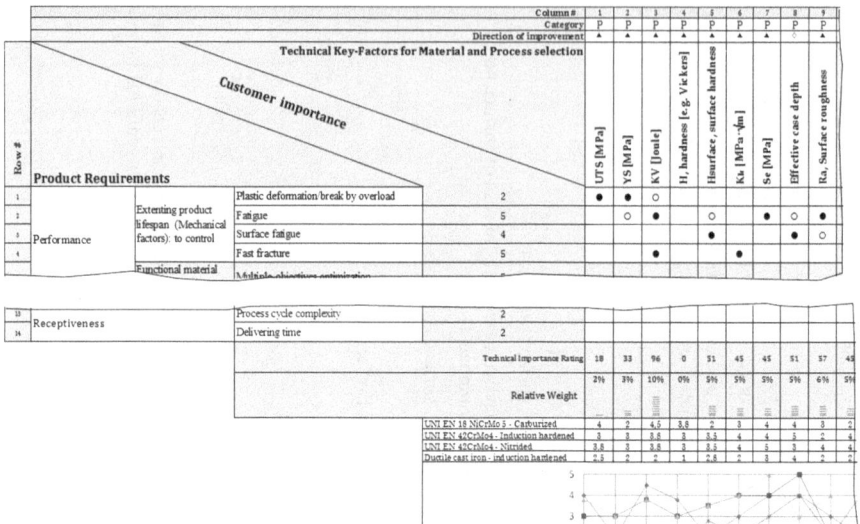

Fig. 3.5 Establishing a link relationship matrix extract from the crankshaft case study

Let us start from the *life span* factor. To this scope, on the basis of their own experience coming from sector literature reviews and some proprietary data information on premature failures provided by the sales business unit, the design engineers of our multidiscipline teams have identified the usual *damage modes* to which a motor crankshaft can be subjected, namely:

- Surface fatigue;
- Fatigue;
- Fast fracture;
- Overloads.

In order to prevent these potential types of damage, Table 3.3 suggests which *Technical Key Factors* of the materials have to be checked (either increasing or decreasing) in order to mitigate damage mechanisms and prevent/postpone possible failures:

- The mechanical static resistance of the base material, expressed mainly in UTS, YS, H;
- The mechanical static resistance properties of the base material, expressed mainly by the hardness of the surface $H_{surface}$ (i.e., top layer surface) and the effective case depth;
- The resistance of the treated surface material to cyclic loads, as expressed by the fatigue limit, Se, for the specific load condition;
- The toughness properties, expressed mainly in KIc, KV, and by the correlated E% parameter (refer to the simplified scheme in Fig. 3.4);
- Surface condition, as in surface roughness.

Table 3.5 Start values for two material performance indices for three materials

	$I_1 = \sigma_f^{1/2}/\rho$ (MPa$^{0.5}$dm^3/kg)	$I_2 = E^{1/3}/\rho$ (MPa$^{1/3}$dm^3/kg)
Magnesium alloy (type ZW30)	6.18	19.65
Aluminum alloy (type AA7050)	5.41	15.04
Steel (type AISI 4140)	2.77	7.55

We already know that the level of correlation between the columns and rows in the QFD framework is causally represented by the symbols (refer to Chap. 2 "The implicit approach using the Quality Function Deployment matrix"):

- "·" for a strong relationship, numerically evaluated as 9;
- "o" for a medium relationship, numerically evaluated as 3;
- "∇" for a weak relationship, numerically evaluated as 1;
- No symbol for a nonexisting relationship, numerically evaluated as 0.

Now we proceed and analyze the *functional objectives and constraints*, the second row, second column in Table 3.2. For this purpose, we can exploit what literature provides. Note our target here: once again we need to assess materials that are optimal for achieving multiple objectives and constraints in product functionality (e.g., stiff and light but strength limited; strong and light at fixed stiffness, etc.). Since this is a tool we already know, which my colleague Ashby has already powerfully provided you with, we are guessing if material indices may be introduced in our matrix. Let us check.

Embedding the Material Performance Indexes and Multiobjective Optimization

What you will be reading in this paragraph is a more technical aspect that explains a procedure which, on the other hand, appears very simple. I do not think that all my readers will be interested in the theoretical basis enabling them to make the very simple move so that they can properly include a *Global Performance Index* in their own *QFD4Mat*, which embeds the different material indices they may be taking into consideration. However, in the following paragraph, if you are curious, you could take a look at what it is happening behind the scenes. You do not have to read it, but you may find it useful if you love demonstrations.

Let us consider an example. Assume you have three materials and you want quickly to express in a single parameter—we have called it the *Global Material Index*—two different performance indicators, I_1 and I_2. The starting situation is illustrated in Table 3.5.

Proceed in the following six steps and do not worry, excel will work for you. You can just follow the instructions below in an excel table or you can download the excel file for this calculation that you can find on the website www.qfd4mat.com. For each material (each row) in the Table 2.5:

- Switch each indicator that has a positive trend in the functions optimization into its reciprocal; in our example, in Table 3.5, both indicators I_1 and I_2 positively contribute in their increase to the escalation of product performance, so we need to convert them into their reciprocal I_{1rec} and I_{2rec}. The result of this first operation is shown in the second box (Table 3.5);
- Convert the I_{1rec} and I_{2rec} of box 2 into a 0–5 scale of values by using the (Eq. 2.14) here rewritten for your convenience:

$$Converted\ Value_{direct} = \left(\frac{Value\ to\ convert}{Average\ Value}\right) \times 3.$$

- The results I_1' and I_2' of this operation are shown in box 3 (Table 3.5)[7];
- For the two indicators I_1 and I_2, define their weight factors (which are the same for the converted values I_1' and I_2'). In our example, we define that I_1 has a weight factor 4 and I_2 has a weight factor 3. The choice of weigh factors depends on what your engineers decide in terms of the relative importance of one objective over the others;
- Multiply the weight factors by I_1' and I_2'; the result of this operation is shown in box 5 (Table 3.5);
- Calculate for each material (each row of box 4) the sum of the logarithmic values of I_1'' and I_2'', applying the $U = \sum_{i=1,n} \log(I_i'')$. The result of this operation is shown in box 6 (Table 3.5). The most optimized solution is researched for the minimum value of U.
- Finally, for each material candidate, convert U into a 0–5 scale of values, P, by using Eq. 2.15). Because we want to set things in such a way that the higher the P is, the higher is the performance of the product, we need to calculate values on a 0–5 scale by using the inverting relation we met in Eq. 3.1, here reported for convenience:

$$Converted\ Value_{Inverse} = 3 - \left(3 \times \left(\frac{Value\ to\ convert}{Average\ Vvalue}\right) - 3\right). \tag{3.1}$$

- The final results are shown in box 7; the global indices P_1, P_2, and P_3 obtained for three candidate materials taking into account the impact of weight factors. The magnesium-based material is the optimal solution.

---

[7] Be careful when you are managing performance indicators with an opposite trend for objective optimization: in that case just the increasing trend indicator will be converted into the reciprocal value.

BOX 1			
Material		$I_1 = \sigma_r^{1/2} / \rho$ (MPa$^{0.5}$dm^3/kg)	$I_2 = E^{1/3} / \rho$ (MPa$^{1/3}$dm^3/kg)
Magnesium alloy (type ZW30)		6.18	19.65
Aluminum alloy (type AA7050)		5.41	15.04
Steel (type AISI 4140)		2.75	7.55

BOX 2			
Material		$I_{1rec} = 1/I_1$	$I_{2rec} = 1/I_2$
Magnesium alloy (type ZW30)		0.16	0.05
Aluminum alloy (type AA7050)		0.18	0.06
Steel (type AISI 4140)		0.37	0.13
	mean	0.23	0.08

BOX 3			
Trend for optimization		↓	↓
Index		I_1'	I_2'
Magnesium alloy (type ZW30)		2.03	1.83
Aluminum alloy (type AA7050)		2.32	2.39
Steel (type AISI 4140)		4.64	4.77

BOX 4			
Index		I_1	I_2
Weigh factor (0-5)		4	3

BOX 5			BOX 6	BOX 7
Trend for optimization	↓	↓	$U = \Sigma \text{LOG} (I_i")$	P
Index	$I_1"$	$I_2"$		
Magnesium alloy (type ZW30)	8.12	5.49	1.65	**3.48**
Aluminum alloy (type AA7050)	9.28	7.18	1.82	3.21
Steel (type AISI 4140)	18.59	14.31	2.42	2.30
		mean	1.96	

Fig. 3.6 The flow chart for the calculation the Global Index P (box 7) for three candidate materials

Once we have calculated the P Global Index for our project, we can finally introduce it into the *QFD4Mat* in order to complete the analysis (Fig. 3.7).

The Global Index in QFD4Mat: Going Behind the Scenes

Suppose for three candidate materials A, B, and C we have calculated the I_1 and I_2 material indices by identifying the two performance equations. Using a simple algebraic operation, we can turn each indicator into a negative trend for product optimization. In other words, we want to have at our disposal all the indicators in such a form that, the lower is their value, the higher is the performance of the component, in line with our objective functions.

Thus, for each material A, B, and C we turn the two indices as calculated into 0–5 on a discrete scale of nondimensional values, I_1' and I_2' (refer to box 3 in Fig. 3.6).

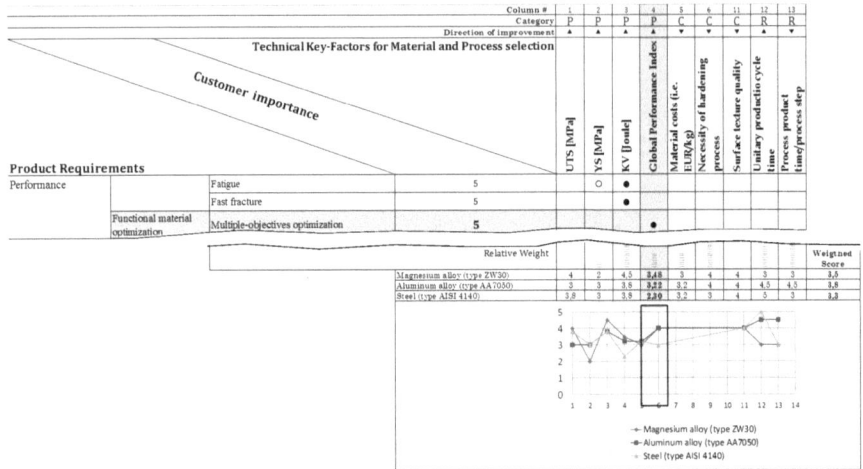

Fig. 3.7 Assembling the global performance index in the *QFD4Mat*

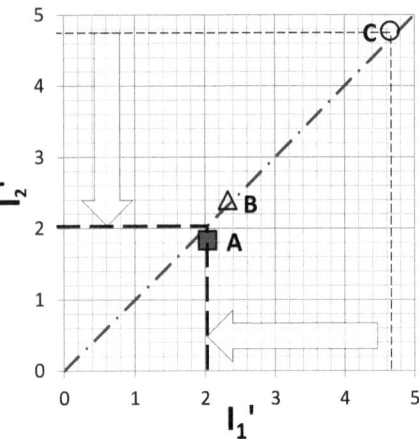

Fig. 3.8 Graphically searching for the optimized solution among the three possibilities A, B, and C

Now we can plot a diagram with I_1' and I_2' as axes, as shown in Fig. 3.8; the $I_1'=I_2'$ line is illustrated. The best solution that will solve our problem of optimization as it is defined so far consists, therefore, in searching for the minimum couple of values I_1' and I_2'. Graphically speaking, the best solution is represented by the point among A, B, and C that is left inside a box-shaped area that shrinks to the bottom left of the diagram enveloping with its corners the $I_1'=I_2'$ line. In our example shown in Fig. 3.8, the point A is the last one left by the shrinking of the box-shaped area.

This is valid as long as the weight of the material indices I_1' and I_2' is the same. How do we proceed in cases where we want to consider two different impacts of the two indices?

Fig. 3.9 Graphically searching for the weighted optimized solution among the three possibilities

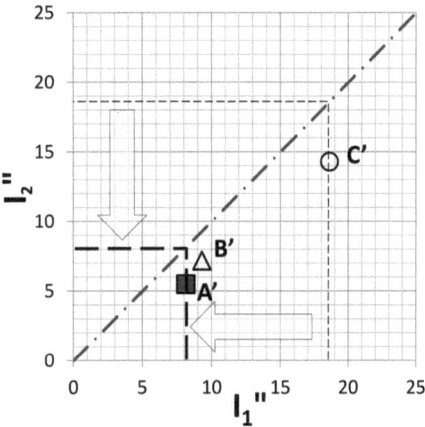

For simplicity, and in order to adhere to QFD approach habits, let us consider we want to weigh two indices differently by using a 0–5 scale factor. Thus, for example, let us assume we want to use 4 and 3 respectively as weight factors for I_1 and I_2. The new weighted indices are obtained by multiplying the weights for the indices, namely $I_1''=4 \cdot I_1'$ and $I_1''=3 \cdot I_1'$. Graphically, this is represented by three new couples of coordinates for the three points A', B', and C' plotted in a 25 to 25 scale diagram (Fig. 3.9).

The procedure used to make a graphic search for the new solution to the optimization problem is the same as the one we have just described above: we plot a box with corners on the $I_1''=I_2''$ line and try to shrink it to the bottom left, until one of the three materials is left inside the box. The material left is the optimal solution for our problem. Therefore, the optimal solution is again that one with the minimum couple of variables I_1 and I_2.

Now let us consider whether we can substitute the box-shaped area we have used with the more flexible *utility curves* that we met in Chap. 2.

By the general *Cobb–Douglas* formulation (refer to Chap. 2), we know such curves are defined by an equation like $u=x^a \cdot y^b$, where u is a constant variable that defines all the possible curves of the family, while a and b parameters define the relative weight of the x and y variables (Fig. 3.10).

We can observe that points A', B', and C' that we plotted in the 25 to 25 scale map in Fig. 3.9, embed the different weights of variables I_1'' and I_2''. This makes sure that if we want to fit points A', B', and C' from the utility family curve, we are allowed to use the simplest symmetric Cobb–Douglas family $u=x \cdot y$ (where a and b parameters have same impact). The graph that results from intercepting points A', B', and C' with the appropriate utility values (i.e., u=46, u=65, and u=265) is shown in the Fig. 3.11.

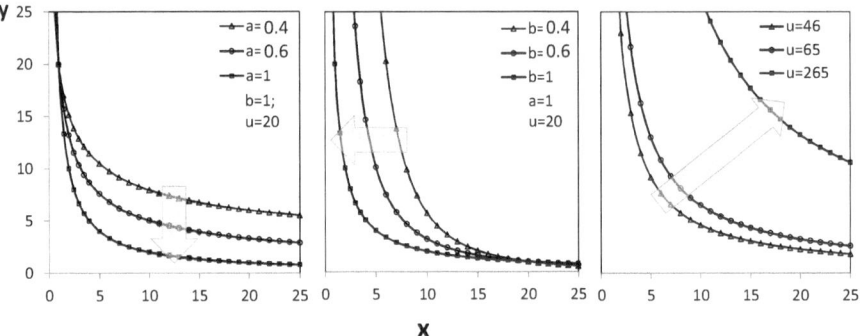

Fig. 3.10 Family curves of Cobb–Douglass in the general formulation $u=x^a \cdot y^b$

Fig. 3.11 Utility curves
$u=46$, $u=65$, and $u=265$ that
intercept respectively points
A, B, and C. The point A
with the lowest utility value
among the candidates has the
higher performance P

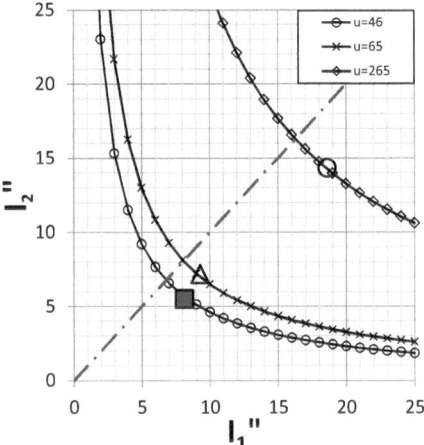

Note that a *Cobb–Douglas* is more often used in its logarithmic form (refer to Eq. 2.8, Chap. 2). For this reason, the utility family curve $u=I_1'' \, I_2''$ plotted in Fig. 3.11 can also be represented in the form:

$$U = \log u \,(x, y) = \log (I_1'') + \log (I_2'').$$ (3.2)

For each candidate material therefore, we can calculate the value U by the above equation.

Once you have calculated the U values, you can always convert them into a nondimensional 0–5 scale. Just keep in mind, however, that the lower the U is, the higher is the global weighted performance P, so in representing the final results it is convenient to use the formula for the inverse conversion (ref. Eq. 3.1, Chap. 2).

The main advantage of using the *Cobb–Douglas* equation in logarithmic form consists in the possibility it gives of introducing more than two variables as those manageable by the graphic solution: in other words, in more complex cases where n performance indices need to be grouped into a global performance index, Eq. 3.2 can be most generally rewritten as:

$$U = \sum_{i=1,n} \log(I_i^{''}).$$

(3.3)

Just remember to apply the algorithm described above as it is fundamental that all indices are inversely proportional to performance: the lower is the index, the higher is the performance.

Construct the Graphic Tool: Completing the QFD4Mat

Figure 3.12 shows a possible matrix for material selection customized for the crank-shaft case study. What emerges is the use of a global performance index that groups together two objective functions aiming to search out the optimal choice for mini-mizing the mass by:

- a stiffness-limited design that is pursued by minimizing the indices $\rho/G^{1/2}$ for the scheme of the shaft in torsion and $\rho E^{1/2}$ for a beam under bending load;
- a strength-limited design, that leads to the index minimizing $\rho\sigma_f^{2/3}$ for both the load case of the shaft in torsion and the beam under bending.[8]

After the main commitment has been carried out by the engineers, to complete the *QFD4Mat* you need high-quality information from your production and sales manager to breakdown the cost category. The whole product process cycle, from the cradle (materials in origin put in the process) to the exit gate (product delivered), needs to be studied to identify costs on which the original choice of materials may impact. In the case study of the crankshaft, costs are explored by organizing a sub-set of the sources of potential, expected, or unexpected expenditure structured into primary, secondary, and tertiary category levels. An example of a framework that would help you for your own product and process cycle is provided in Table 3.6, extracted from *QFD4Mat* in Fig. 3.12.

The last effort to be made is to complete the analysis of the technical require-ments and their relative correlated material key features by exploring the vast class of product requirements that are grouped in the receptiveness category. If we want to put it in a nutshell, we can say that this class of attributes deals with the capability of materials to influence in any manner the positive response of our customers. We just need to remind ourselves that the customer for our product is not restricted to the user, but to each part of the product value chain to which we contribute in order to achieve that final consumer. The receptiveness requirements you can explore in

---

[8] *Ibid.* Ashby 2005.

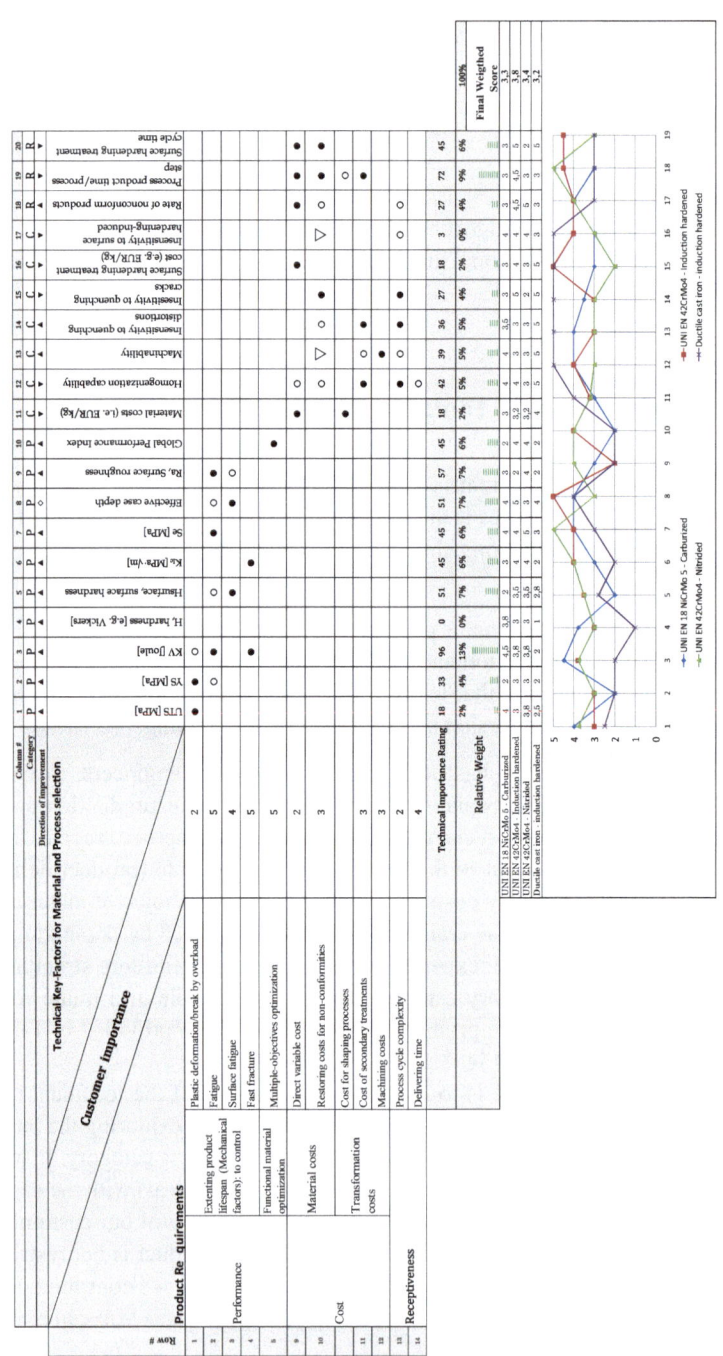

Fig.3.12 The customized *QFD4Mat* for the crankshaft case study

Table 3.6 Product requirements for the crankshaft case in Chap. 1; the "cost" attribute class is organized by the usual QFD structure into primary, secondary, and tertiary category levels

Primary	Secondary	Tertiary
Cost	Material costs	Direct variable cost
		Restoring costs for nonconformities
	Transformation costs	Cost for shaping processes
		Cost of secondary treatments
		Machining costs

the crankshaft example are very limited, but this is not so in many products that have a direct interaction with consumers.

In these cases, the product material features serve not only to increase performance and control cost, but they can directly boost the user's perception of the benefits he or she is receiving, namely the value that the customer derives from the product because of the materials themselves. The laptop I am typing on is an experience for me, because of the visual and tactile attributes given by the materials under my fingers. The satin finished aluminum suggests quality to me and durability; the metal used for the case where my hands lie make me feel it is "cold" to my touch. A material feels "cold" to the touch if it conducts heat away from the finger quickly, and this depends on its thermal conductivity and on its specific heat Cp. And we can introduce such "coldness to touch" in the receptiveness requirements and correlate it with the quantity IρCpλ[9]

On the opposite scale, my friend a dentist, who is passionate about materials, asks for dental instruments that can reduce hand fatigue, which are soft to the touch but ergonomically designed to be precise in their use. The "softness but precise to the touch" requirement can therefore be expressed by a material attribute that combines both the elastic modulus and hardness, translated by the key features E·and H, where E is the Young Modulus and H is the hardness.

The surface texture we experience when looking at and touching a product is an attribute an industrial designer seeks in order to achieve product recognition and appeal, working alongside an engineer whose job it is to enhance functionality. It may be the fact that I am a metallurgist that led me to turn away from cheaper plastic and search out greater freedom of form and color for a simple shape made of satin aluminum. But my fingers and my eyes appreciate the choice that I have made. And as deeper explained in the following chapter, this is typical of what your industrial designers in the R&D unit would call the "emotional appeal" of products that express their own "personality" through materials.

---

[9] *Ibid.* Ashby 2005.

Chapter 4
The Use of QFD4Mat and Graphic Tools in Product Development Processes

Abstract **Chapter 4** starts by outlining the new product design process in brief, then moving toward how it is possible to use the QFD4Mat tool in order to illustrate the ways in which an efficient design proposal can be generated in the conceptual design phase. In order to aid the reader with a non-technical background in assessing the final results obtained by the QFD4Mat, this chapter also presents two graphic analysis tools, one that is well-known as the "value curve" of the product, the other an unpublished original Bubble Map tool that has been fine-tuned by intense classroom work in last three years.

Introduction

A variety of books deal with new product development (NPD) and product innovation issues, which is the most ambitious fruition of NPD, focusing on a multiplicity of aspects. Some books focus on the early stages of the idea generation process, some others on how consumers' needs can effectively be addressed for the success of the products on the market, some devote most of their attention to the management aspects of product development, while others illustrate how organizations put engineering processes in action in order to translate ideas into products to be delivered and sold.

A common question seems, however, to arise from each point of view: Should an innovation process start from the market demand or should it try to stimulate, offering new technological solutions? Local markets today are evolving into something more sophisticated, demanding, and well informed. The result is that market competition will constantly intensify. A further complication in more recent times is the increasingly aggressive competition in search of the product with the lowest cost. But, will understanding the evolution of the model for product innovation that we have experimented so far help us to understand what to do today in order to be successful tomorrow? Regarding this question, there is usually a face-off between two approaches as can happen paradoxically if you analyze the real nature of the same enterprise.

Some people believe that the success of enterprises is to be interpreted through the lens of the technological research they carry out, the fact that they are able to

© Springer International Publishing Switzerland 2015   99
F. D'Errico, *Material Selections by a Hybrid Multi-Criteria Approach,*
SpringerBriefs in Materials, DOI 10.1007/978-3-319-13030-9_4

acquire talent and then create the environment in which this talent can reach its maximum expression. There are those who, on the contrary, interpret the success of products as something that comes about because they are developed on the basis of market expectations, which are often latent in product users. The latter concept is a bit obscure, but it can also be translated into: "think about the new product as if you were a fascinated consumer, freeing out every possible idea so that you can turn the current product into the very thing I would like it to be." A competitive market generally results in users that demand products that are able to give complete satisfaction for the money spent. This does not always mean making the product at the lowest possible cost, thus putting it onto marketplace at the lowest price. In other words, the price is not the only factor in the business of choosing today since, as consumers, we all make decisions keeping several choices in mind: "I might choose this, but it must suit me." The benefits should be intelligible, since the comparison between alternative offers can be made today with the touch of a finger on our pocket smartphones. For producers, the global market means the *competitive arena* where they challenge each other; for the consumers it means an instantaneous global shop window. Thus, which of the two interpreters of product success is the one that has got it right? Both of them, but sometimes the two still communicate a little with each other. When a fusion of points of views occurs, the result far exceeds the sum of the two. An example of this was given some time ago when Jobs asked his engineers to develop a mobile phone with a single button. They replied "Foolish!," but the challenge had already been launched.

The New Product Design (NPD) Process in Brief

Booz et al. (1982) have proposed a systematic model for the development of a new product based on eight phases: new product strategy, idea generation, screening, assessment, analysis concept test of business, development, testing, and marketing. But several improved versions of this model for product development have occurred in the years since World War II. We will now take a brief look at these.

During the first 20 years after World War II, an extensive supply context was developed: very emergent new industries that held new technologies (often derived from the huge advancements in technology achieved during the World conflict) were largely supported by government policies, many supply-pushed to provide technological goods into millions of citizens' houses. New stimuli were put into scientific advancements in universities and government laboratories and in the development of skilled man power. The commercialization of such technological changes to families was generally perceived as a natural and unstoppable pathway from scientific discovery, through technological development in firms, to the marketplace.

Toward the second half of the 1960s till the early 1970s, a general prosperity remained high since in many countries manufacturing continued to grow along with productivity, while employment was still increasing, even though at a much reduced rate (Rothwell and Soete 1983). The more efficient companies were now starting to fight for their market share; product development strategies (i.e., analyses of what decisions we should take today in order to create the sales we want tomorrow) developed and introduced some of the basic concepts of contemporary marketing. Product innovation processes embedded the market emphasis on "demand side" factors, the market place made of consumers. This was the rise of the second generation of the NPD process, also called the "market-pull" (sometimes referred to as the "need-pull"); it was a model of product innovation processes where the market is the fundamental source of ideas. But it was occurring in a period when resources were (still) relatively abundant.

The context concerning the availability of resources radically changed in the early to late 1970s, when two major oil crises determined a prolonged period of high rates of inflation and demand saturation. This was a critical context that spelled out a cataclysm for enterprises that had a high production volume; supply capacity generally exceeded demand and led to growing structural unemployment. The rationalization of sources and resources was a strategy followed by most companies. The structure of successful innovation was deeply researched (Cooper 1994). Empirical results indicated that technology-push and need-pull models of innovation were extreme and atypical examples of a more general process of interaction (Mowery and Rosenberg 1978). The third generation of innovation is interactive, namely a sequential process with several feedback loops.

The early 1980s was a period of economic recovery, and the notion of global strategy emerged. The speed of development of new products became a determining factor in competition, so that firms adopted time-to-market-based strategies. The Japanese were more powerful and rapid innovators than their Western counterparts. These features still live today, since speed and efficiency are fundamentals in the fifth generation of product innovation processes, whose success derives from the efficient and real time handling of information across the whole system of innovation (from internal functions to suppliers, customers, and collaborators).

Five general characteristics have so far been found to be dominant for the adoption rate of innovative solutions, namely innovative products[1], in the marketplace (Rogers 2003):

- *Relative advantage*: is the degree to which an innovation is perceived to be better than the product it supersedes, or than competing products; it can be measured in economic terms, but social prestige, convenience, and satisfaction factors are also relevant. The perception of advantages instead of a real objective advantage is the most important factor that can speed up the rate of adoption of innovative products;
- *Compatibility*: is the degree to which an innovation is perceived to be consistent with the existing values and norms of the social systems in which we wish to distribute;

---

[1] Actually, the analysis that follows, it is not focused on product innovation, as the general principles are also valid for services and processes—namely ideas in general.

- *Complexity*: is the degree to which an innovation is perceived to be difficult to understand or use; new ideas translated by easy-to-understand, user-friendly products will be distributed and adopted more rapidly than innovations that require new skills and understandings;
- *Trialability*: is the degree to which an innovation can be experimented with on a limited basis. An innovation that is *trialable* represents less uncertainty to potential adopters and allows for learning by doing. Innovations that can be trialed will generally be adopted more quickly than those which cannot (think for example to new growing market of Apps for the smartphone);
- *Observability*: is the degree to which the results of an innovation are visible to others. The easier it is for others to see the benefits of an innovation, the more likelihood there is it will be adopted. Since visibility stimulates discussion on the new solutions and ideas brought on the scene by a new product as it enters a society, a sort of epidemic model of diffusion spreads it to potential adopters.

And at the origin of the circulation process of new products that strive to become a market success, there is the innovation development process. It consists in all the decisions and activities starting from the identification of product ideas and the recognition of the key-features in a need or a problem, then following through research and elaboration to the completion of a new product development cycle with the launching, commercialization, diffusion and adoption (above depicted by five main forces), and the impacts and consequences of that adoption. A quick schematic snapshot of the new product life-cycle process and its implications on company business is outlined in Fig. 4.1. As usual, costs start from the beginning of research and development since they are generated at least among internal units (departments) of the company or with external technology suppliers.

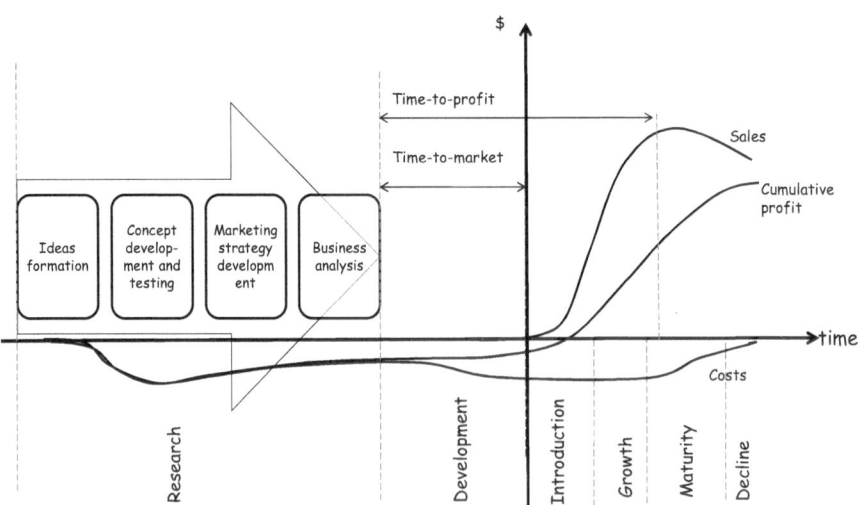

Fig. 4.1 General timeline of an NPD process with its natural life-cycle onto marketplace

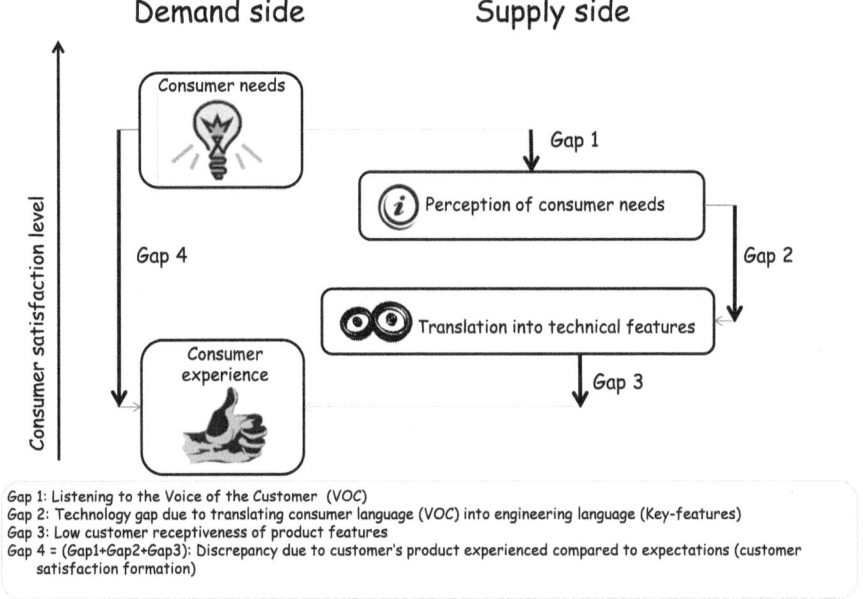

Fig. 4.2 Simplified gap model for a new tangible product

The common belief is that a company that leads and develops new products must anticipate the problems of (targeted) individuals, who will be the ultimate adopters of that product. But a robust information-exchange network is necessary inside the company so as to reduce the gap between the actual answers that the product wishes to provide and the users' real needs, for this is a key issue influencing the NPD process. Figure 4.2 illustrates in a simplified model, an adaptation of the *Gaps Model of Service Quality* (Parasuraman et al. 1991), how the evaluation of product performance forms into adopters.

The main effort the NPD team faces is acquiring and exchanging data about the perceived performance offered by the commercialized solution—*compatible* with existing values and norms—which will in its turn provide rapid diffusion due to the product's low *complexity* and properly communicate its *relative advantage* over the competitors by *observable* and *trialable* features.

A company's NPD team is, therefore, committed to the responsibility of taking decisions that will impact over various phases of the project (from the generation of the initial idea right up to the final commercialization of the product) working in a context with a high degree of uncertainty. Thus, the multidisciplinary skills required should restrain these uncertainties connected with multifaceted problems, which sometimes may require continuous negotiation concerning the designers' creativity and imagination, the marketing sector's ability to detect product needs from within the marketplace, the engineers' ability to transform creativity into features and functions that are fully perceived by users and the financial department, who procure resources and plan and assess the NPD project sustainability over the product life

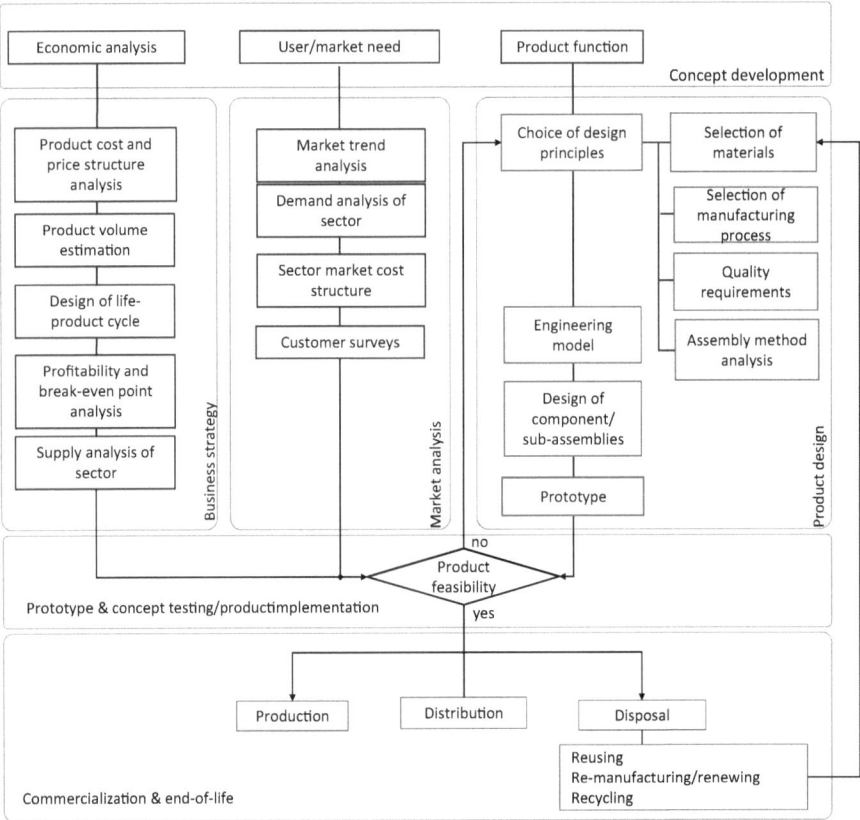

Fig. 4.3 General workflow scheme involving various phases and competences in an organization illustrating internal transfunctional work processes during an NPD process

cycle (see again Fig. 4.1). A "white-box" flow of information as that one depicted in Fig. 4.3 is the key-process that can form, make grow, and change the insights playing a role in product functionality, performance, easiness of assembly, maintenance and durability, capability for recycling, and make all these features interact with cost estimation and manufacturability (Ulrich 2011; Mital 2014).

A new product that aspires to be an opportunity to create value through further investment requires more than an intuition if it is to become a market success. It takes a cross-functional team capable of the in-depth evaluation of several features that can impact on the product as a whole once it has been designed and which will satisfy its customers when it has been produced, delivered, and sold.

Summarizing, what we can state is that an innovative product will be successful in the market if the consumer:

- can *recognize* its innovative aspects/functions, even in cases it is less cheap (relative advantages);

- can *comprehend* that those relative advantages match his standard expectations (receptiveness).

A major question is, therefore, how to assess in the early conceptual design phase the product key features that are taking minimum or acceptable risks. In Chap. 3, we acquired a lot of precious information from our skilled colleagues on the product itself, and now we need to summarize them to check at the early stage of the NPD process that we are on the right track. But we need to do it in a quick and simple way that can be comprehended by everyone sitting around our table.

How to Generate an Efficient Design Proposal in the Conceptual Design Phase

A friend of mine is an artist. I like his art and the way he expresses himself. One curious question I asked him (the same that I guess many of you, if you ever talk frankly with an artist, would ask) was how can he transform an idea, or an image that comes up in his mind, into a painting that is not just pretty, but can be recognized by people as an artistic expression of emotions. "It's just a miscellany of good painting techniques and something of your own that people recognize as new and innovative," was his answer. My friend has a very good technique for his figurative paintings, which have a high definition of the human body. He uses only two opposite colors, black and white. No other colors at all. And he makes all the paintings without any brushes, just using his fingers (Fig. 4.4).

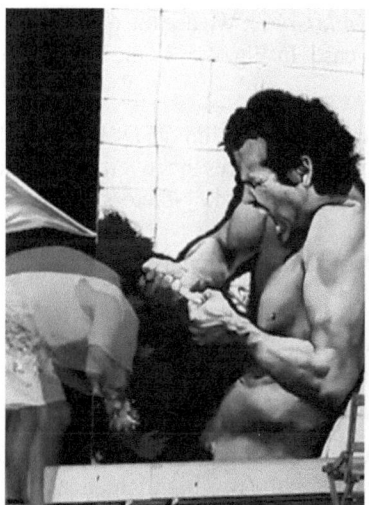

Fig. 4.4 Making of an artistic hand painting. (Courtesy of Paolo Troilo)

So I said to myself: "making an artistic painting is just like a NPD process!," but I preferred not to share my thought with him; his paintings are technically appreciable and they use essential "raw" materials and resources: two colors and a canvas. And they target their own *functional objective*: to stimulate emotions. They manage, in other words, using no more than is completely indispensable, to achieve the power of condensing a great deal of information—most of it personal and not fully accessible in any other way—in one single "image." I realized this is what we need for completing the path we set off on in Chap. 3: a rapid visualization, just an "image," that can summarize a lot of information which we acquire by the *QFD4Mat*.

Among the several ways of visualizing data, we need something that maintains the high quality of information provided by an image, which can be intuitively comprehensible without requiring any specific skills. Two methods of visualization will, therefore, be illustrated here that can reply to such needs.

The first method we borrow from what Kim and Mauborgne, two management strategy scholars, recently proposed, which they call the *Value Curve* of products, services, and processes (Kim and Mauborgne 1997, 2005). Here we focus specifically on products.

A *Value Curve* is a diagram used to compare products over a range of factors by rating them on a qualitative scale, from low to high. Such factors can be product features, product benefits, or ways in which a product is distributed or consumed. Here we are interested in the product features related to materials. The combination of these various features defines a product, distinguishing it from the other competitors that may have different *Value Curves*. Multiple *Value Curves* can be drawn and superimposed to create a very user-friendly visual comparison among competitive products and to unearth possible gaps in the market. By investigating the feasibility of these gaps, Kim and Mauborgne warn, it may be possible to identify changes to the product that significantly alter the value proposition and enhance the receptiveness of users.

Let us proceed in creating our product *Value Curve*. We use the *QFD4Mat* results we obtained in Chap. 3 for the crankshaft case. In Fig. 4.5, in the next three areas highlighted, we can recognize:

- Area 1 contains the 20 material key-factors identified in *QFD4Mat* as influencing the desired product requirements (the *VOC* features in the left side of matrix);
- Area 2 contains the relative weights of 20 material key-factors, in percentage; and
- Area 3 contains the assessment by a 0–5 nondimensional scale of the material candidates through the 20 key-features.

Simply reorganizing the data in Area 1, Area 2, and Area 3 we can draw our *Value Curves* for 4 materials. Let us do it.

We can proceed as follows:

- Organize the 20 columns of the key-features (i.e., Area 1) by horizontally sorting them, ascending from lower to higher values in the row highlighted in Area 2; the data in Area 3 will be sorted accordingly;
- Plot a diagram with the y-axis ranging between 0 to 5 value and the 20 key-features on x-axis;

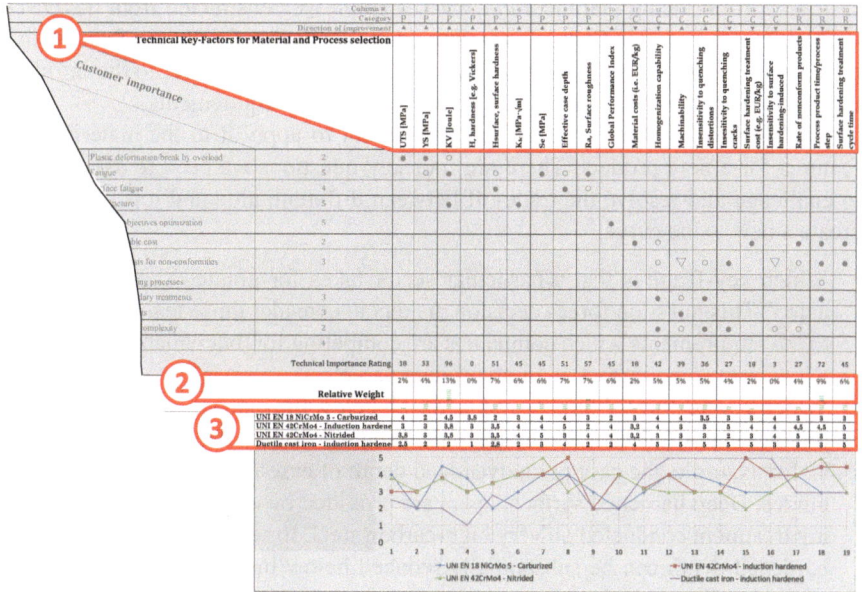

Fig. 4.5 Relevant portions of *QFD4Mat* that has been constructed for the case study in Chap. 3

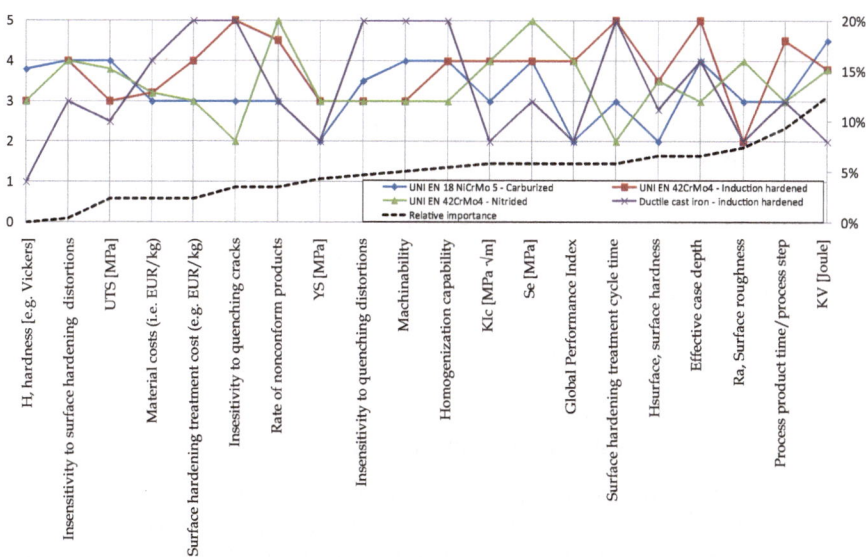

Fig. 4.6 The Value Curve visualizing the final results of *QFD4Mat* in the crankshaft case study, referred to in Chap. 3

- Plot onto such a diagram the data on the 4 materials contained in Area 3.

If you follow the three-step procedure listed above, the result is that shown in Fig. 4.6—the *Value Curves* of the 4 candidate materials along with the 20 key-features organized from the least important (on the left) to the higher relative importance (to the right) as also illustrated by the dotted line "Relative Importance."

It is worth noticing that the *Value Curves* graph we constructed from the *QFD-4Mat* data elaborated in Chap. 3 is an immediate snapshot of such key-features that have been researched and then prioritized in order to get an answer to the information in the left-hand side matrix, the *Voice of Customer* requirements.

To fully exploit *Value Curve* theory, we want to apply it to the material key-features for a new product. There are four key questions we need to ask in order to challenge and resolve the trade-off between differentiation and low cost and to create a new value curve:

- Which key-features that a particular sector takes for granted should be eliminated? This question forces decision makers to consider those features that reply to the requirements which companies are competing for, but which actually have a low or even no impact at all on product value;
- Which key-features should be reduced well below the industry's standard? Some product features do not add value to the product, or they are overdesigned. This is, for example, the case of an oversized depth of case-hardened layers on linear guides. Deep hardened surface layers are provided by critical induction hardening treatment conducted on very high-carbon steel. In several cases, the effective hardened depth can be substantially reduced below the usual sector standards, and induction hardening treatment can be substituted by a less critical gas nitriding process;
- Which key-features should be raised well above the industry's standard? Quality appearance, image, and impact resistance by use of light metal alloys in electronics, for example laptop covers, are today made of machined aluminum instead of the cheaper injection molding polymers, which give greater formability;
- Which key-features should be created, responding to unexplored product requirements for the sector, which the industry has never offered? This forces us to seek out solutions that break out of the industry's sector boundaries, to explore new offer contexts. In the past, but also in contemporary architecture, resistance to the environment, formability, and strength as regards wide panels that are both thin and light led to the introduction in buildings of stainless steel to create value. A historic example is the Chrysler Building in New York, one of the 20th century's architectural trophies. A classic embodiment of the Art Deco style with its distinctive top ornamentations of austenitic stainless steel, which could be free-modeled with features used on Chrysler automobiles. It was the first large, high-profile stainless steel application in the world.

As stated, no tool you use will do the job for you, that is, it does not whisper the solution that will be successful into your ears. But a tool can help you in acquiring high-quality, multifaceted information, which is what the *QFD4Mat* aims to do, and to condense a huge quantity of information codified in a nondimensional 0–5 scale into a visual image that allows you to group technical key-features all together. The tool will do the first part of the job—creativity in solutions is up to you.

Positioning on Market: Customer-Oriented Supplying Using PCR Bubbles Graphic Visualizations

The *Value Curves* visualizing tool is simple and quick to construct. But it is not the only way to summarize the *QFD4Mat* results with a high-quality information content. The second tool presented as follows sets out to be as simple for any user as the *Value Curves*, and it can overcome one limit of the *Value Curves* graph. Let us focus firstly on that very limit.

With a little insight you might have observed in the final passage of the previous paragraph how we have underlined that the *Value Curve* can summarize and visualize competition on the key-features that are considered relevant in that they impact on product requirements. Specifically in our example, we deal with 20 key-features that have been selected on top of the *QFD4Mat* matrix, which aim to answer the 12 product requirements, the left shoulder of the matrix. The *Value Curve* is finally elaborated considering the "importance mix" resulting from the "causal links" established between technical *key-features* (top roof of *QFD4Mat* matrix) and product requirements (or *VOC*, the left shoulder of *QFD4Mat* matrix).

Now we return to the diagram of the simplified *Gap model* shown in the Fig. 4.2; refer particularly to "Gap 1" and "Gap 2": they are respectively those misalignments that are given form respectively during the translation of the product requirements as expressed in the voice of the customer and thus by the identification of such material *key-features* that are thought to be capable of determining the final positive experience of users over and above their original product expectations.

Having been constructed by ordering the assessment of *technical key-features* for material candidates, the *Value Curves* embed possible misalignments between *VOC* and the final user experience of key-features that try to translate *VOC* into a tangible product. In other words, they embed gap 1 and gap 2 in the *Value Curves* creation. If we want to reduce such misalignments, we can therefore use another visualizing tool that we call the "*Performance-Cost-Receptiveness Customer-Oriented Diagrams*" or in a more friendly expression "*PCR bubbles maps*." These are shown below. Firstly, however, we need to understand the origin of the *PCR bubbles maps* diagram.

We start again from the assumptions regarding innovation drivers and the present market context: the challenge today for companies is to compete by acknowledging customers preferences and perspectives, sometimes being able to jump over and perceive what is a not yet clearly expressed as a need. The main scope of strategy analysis for starting an NPD should, therefore, be to map the competitors' relative position as it responds to customers' actual and, if possible, latent needs. In such cases, be warned that the semiquantitative benchmarking of key-features can result in an intrinsically high risk of "Gap1-to-Gap 2" type misalignments (see again Fig. 4.3). When the precious high-quality information in *VOC* is "contaminated," it is therefore advisable to visualize the *VOC*, that is, the user needs directly.

Let us construct our model and then our visualization graphics. At this point, we have all the data information at our disposal: once again, as for the *Value Curves* construction, all that we need has been provided by the *QFD4Mat* tool we used for our case study in Chap. 3.

Before proceeding with the construction of our graphic tool, the next section is devoted to a general comprehension of how we should group information from the *QFD4Mat*. This is not fundamental, but if you are curious about the origin of *PCR Bubble Maps* you could take a look.

The PCR Bubble Map: Origin in Brief

Product character[2] is the result of important features, particularly those that differentiate the product. Around the product scope and its main function, the character of a product can be dissected into a subset of feature categories (Ashby 2005):

- the *context* defines the intentions or "mood" by exploring possible answers to the 5 "*Wh*" questions (Who? Where? When? Why? What?) aiming to define the context and habits of product usage;
- the *materials* and *processes* create the architecture, the hardware of the product;
- the *usability* determines the interface with the user;
- the *personality* of the product, addressed by the aesthetics, associations and perceptions that the product creates in the user's emotions (mainly architectural, like shape, surface finishing, opacity or transparency of surface, colors, etc.).

Whatever category they belong to, the product features contribute to building up the resulting product character. They can be reorganized into the three main categories that we have already encountered in Chap. 3, once again set out here:

- *Performance or P*, which includes all the features related to technical and functional issues that determine the relative advantage of the product technology, the capability to translate ideas into tangible tasks and actions;
- *Cost or C*, which includes all aspects related to the business and economic aspects of the product;
- *Receptiveness or R*, which groups all those features more related to "Product Psychology"[3] (e.g., usability, trialability, complexity, emotions, image, etc.).

Depending on the "side" from which we want to study these product features, namely:

---

[2] Similarly designers talk about the concept of product language, that is defined as the product's ability to communicate information about itself actively in the market by product language: a very heterogeneous form of expression communicated through dimension, form, structure of the physical surface, movement, quality of materials, functions, colors, surface and sounds, all of which have a strong effect—positive or negative—on potential buyers (Gross 1984).

[3] A sharp definition gathers the material and processes categories into a broader class of features that is called "Product Physiology", while the usability and personality categories, namely those features that have the closest relationships with users, are merged into the class called "Product Psychology" (Ashby 2005).

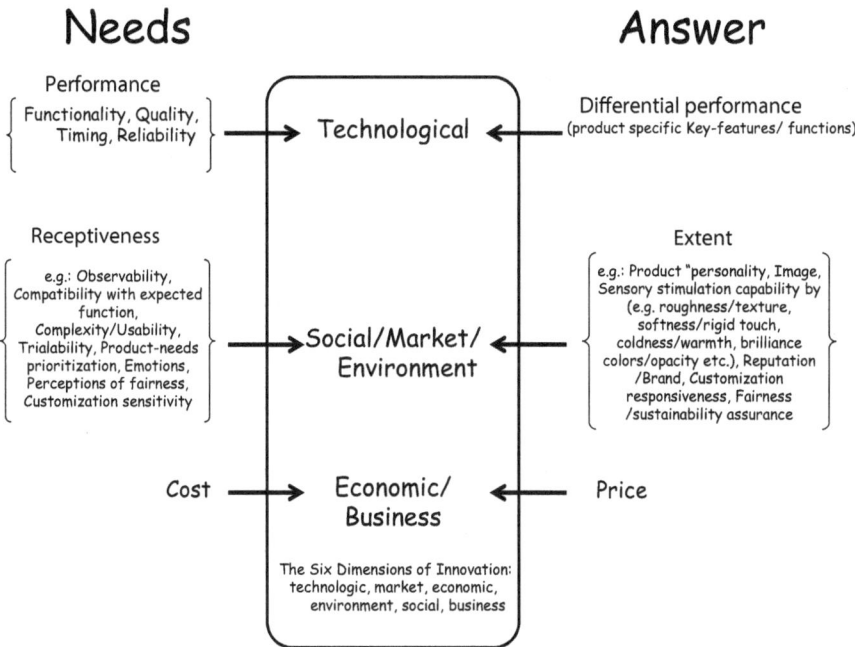

Fig. 4.7 Scheme of model to represent as regards the demand and supply sides all product requirements and product features developed to answer to requirements expressed by *VOC*

a. the side of *demand*, pertaining to choices and preferences (needs), experience and the final comparison with the expectations (customer satisfaction) of consumers;
b. the side of *supply*, pertaining to research through consumer needs and their translation/codification into tangible products;

the three category P, C, and R address respectively: (a) *product requirements* and (b) *product features*; the latter aims to supply the best answers to the fully expressed or latent questions in the former. A general scheme for representing the above PCR model is shown in Fig. 4.7.

We learned in Chap. 2 that when using semiquantitative methods, like a QFD matrix, any requirements and features—measurable or not—can be converted into a simplified 0 to 5 scale (refer to Chap. 2, Eq. 2.14 and 2.15).

We can, therefore, gather the nondimensional 0–5 values obtained for *product requirements* (demand side) and *product features* (supply side) into P, C, and R parameters. As a last task, we therefore plot onto the same diagram for the *supply side* and the *demand side,* two circles with their centers of coordinates, respectively, P and C, and with their diameters R. Since a drawing is better than words, Fig. 4.8 shows the two resulting *bubbles*. The ideal situation is obviously when the *PCR bubble* representing the *supply-side* completely matches the *demand-side PCR bubble.*

Fig. 4.8 Construction of PCR bubbles for the supply side and the demand side

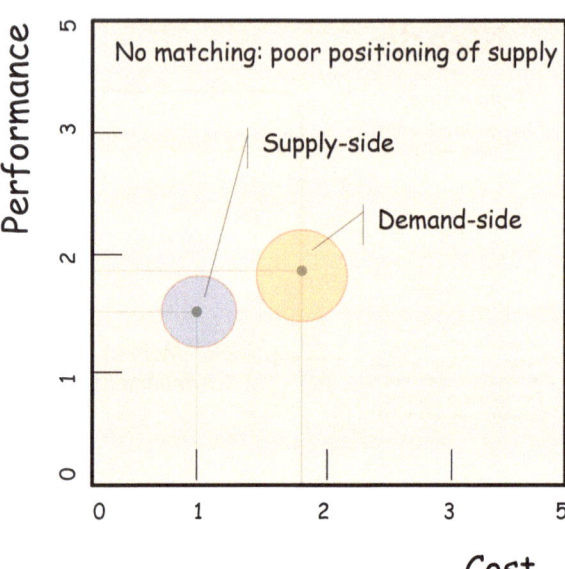

In the graph in Fig. 4.8, for example, the two circles do not match; in this simplified example, this depends on:

- Too great a "distance" existing between the "center coordinates," namely the fact that the positioning of P and R features evaluated onto the product features supplied by the company is too far away from the P and C requirements on the product supplied by users;
- The R parameter is too low. This is the diameter of the circles that graphically explains on a simplified scale: (a) how receptive the consumer is to the requirements of the specific product; (b) to what extent the designed product features "cover" the product requirements expressed by the consumer.

Figure 4.9 shows how the rise of the R parameter can affect the potential matching of the two bubbles. Instead, Fig. 4.10 shows the *supply-side PCR bubble* that partially covers the demand-side bubble. A relative portion of the consumers' bubble schematically (remember that it is a graph constructed in a semiqualitative scale) represents consumer's needs expressed in the form of product requirements. This bubble we call the *Voice of Customers' (VOC)* bubble. When the *VOC* bubble is partially covered by the material key-features' bubble, it means the product value that the suppliers propose partially fits in with the consumers' expectations.

Figure 4.11 summarizes various situations in which the supplying bubbles are positioned outside the demand bubble, but with different results in terms of NPD process results.

Finally, Fig. 4.12 represents a graphic visualization of a benchmarking analysis of market position conducted by the use of PCR bubbles.

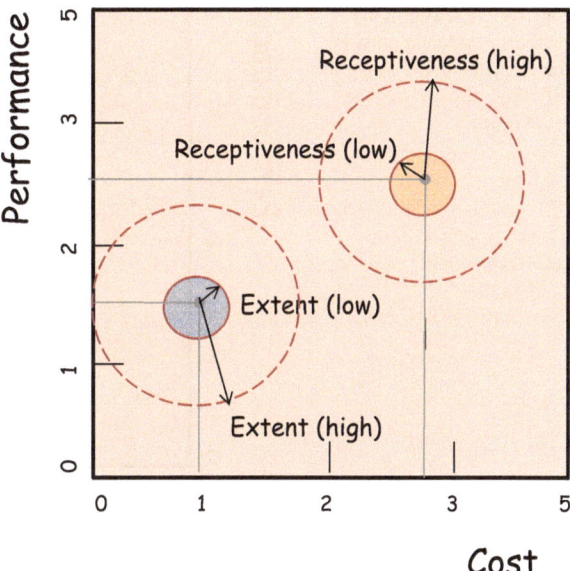

Fig. 4.9 P and C coordinates are fixed, R parameter increases

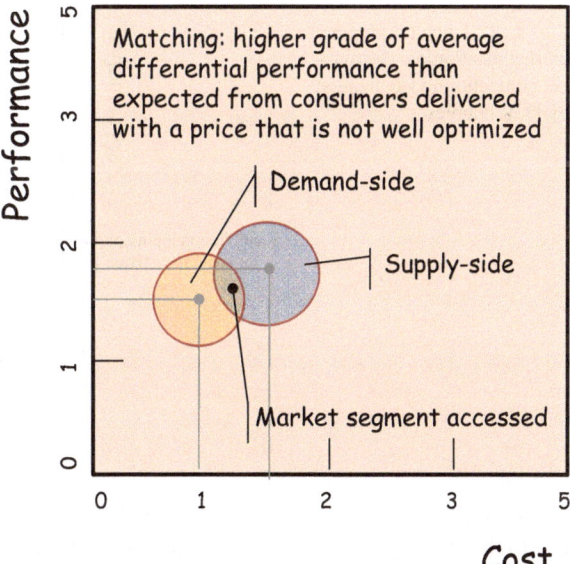

Fig. 4.10 The two PCR bubbles match: the product feature proposal regarding supply partially covers the demand expectations

A Case Study: The QFD4Mat Results by Bubbles Visualization

In the following, we will construct the *PCR bubbles map* for the case study we faced in Chap. 3 (refer to Fig. 3.12) by using all the data results we obtained from the construction of the *QFD4Mat* for the classroom case study (also downloadable at Web

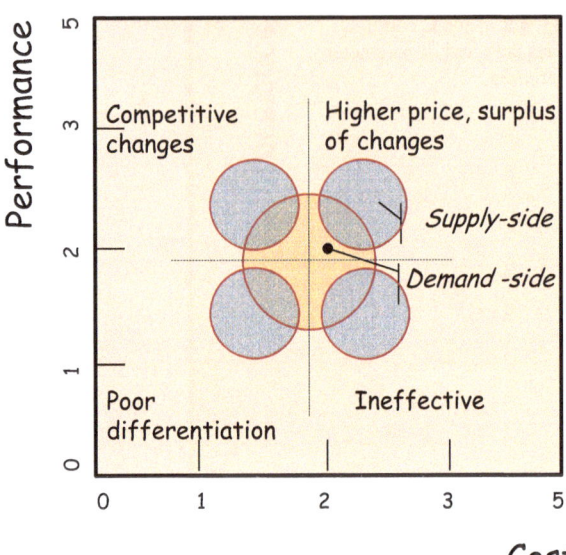

Fig. 4.11 The supplying bubbles positioned outside the demand bubble imply the different efficiency of the NPD process

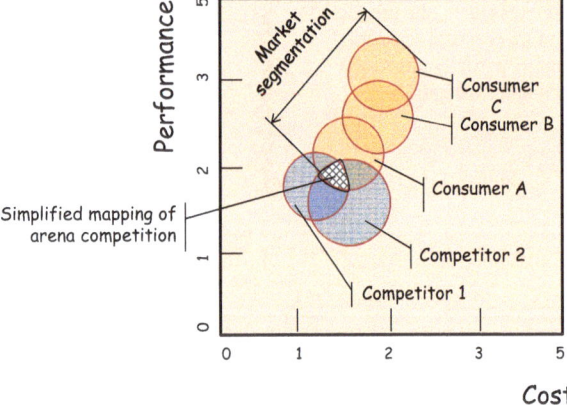

Fig. 4.12 Benchmarking and market positioning: the market competition visualization by PCR bubbles

site *www.qfd4mat.com*). Here below, Fig. 4.13 shows the same matrix as Fig. 3.12 but highlights 4 box-shaped areas: they are grouped in two couples of two boxes that are linked by "A" and "B" pathways.

The two boxes linked by the "A" connection are located in the left shoulder of *QFD4Mat* matrix, namely where we have gathered the *product requirements* expressed by the *voice of the Customer (VOC)*. Particularly, the second box (from left to right) linked by the "A" connection contains in rows the relative importance values of the 12 product requirements, expressed by the usual for us 0–5 scale. We can reorganize these data as in Table 4.1: for each group of product requirements that refers to P, C, and R parameters, we calculate the mean value, as it is reported in the last column of Table 4.1. This simple operation allows us to express the *VOC*

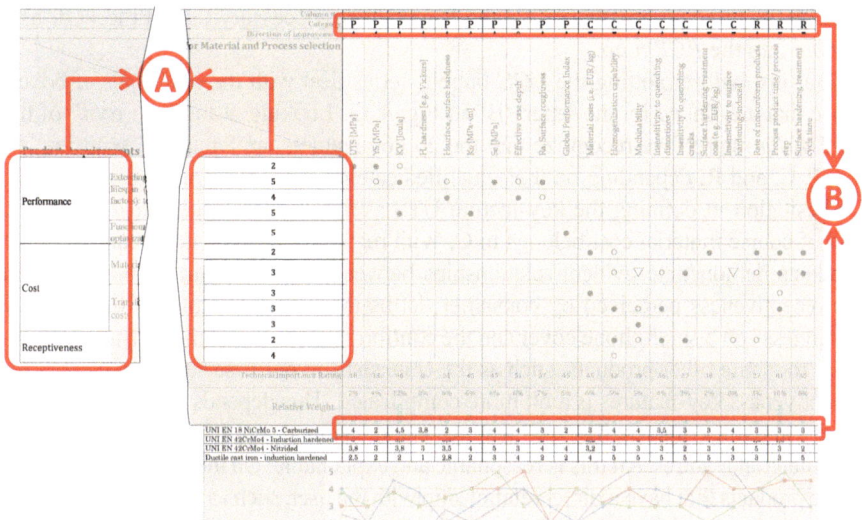

Fig. 4.13 Matrix QFD4Mat of Fig. 3.12, Chap. 3. There are 4 highlighted boxes: two pairs of boxes are considered together as letter *A* and *B*

Table 4.1 Data extracted from the *QFD4Mat* of Fig. 3.12: the "left shoulder" data have been organized for the calculation of the mean values for the P-type, C-type, and R-type product requirements

Primary	Secondary	Tertiary	Assigned value	Mean
Performance	Extending product lifespan (Mechanical factors) to control	Plastic deformation/break by overload	2.0	4.2
		Fatigue	5.0	
		Surface fatigue	4.0	
		Fast fracture	5.0	
	Functional material optimization	Multiple-objectives optimization	5.0	
Cost	Material costs	Direct variable cost	2.0	2.2
		Restoring costs for nonconformities	3.0	
	Transformation costs	Cost for shaping processes	3.0	
		Cost of secondary treatments	3.0	
		Machining costs	3.0	
Receptiveness		Process cycle complexity	2.0	3.0
		Delivering time	4.0	

data by the three parameters, P, C, and R, which are necessary to plot the *VOC* as a *bubble* into the *PCR map*.

But just before plotting the *VOC* bubble, let us deal with the other pair of boxes, those that are linked by the letter "B" in Fig. 4.13. Looking at the "top roof" of the *QFD4Mat* matrix, we recognize that the 20 key-features are "marked" with the letters P, C, and R, depending on their qualified category. For example, we would have no hesitation in defining the key-feature UTS as a feature belonging to *P* category; and of course material cost belongs to C. With regard to receptiveness features, some products demonstrate evident relationships between the key-features addressed and the receptiveness perceived by consumer; let us think, for example, of the "surface roughness" on a smartphone cover and the comfort to the touch of the consumer. However, for some other products, such as the crankshaft in our case study, the search for receptiveness key-features is a bit more complicated. This depends on a limited availability in reality of such "R-type" features—and can be true for both *VOC* and in the key-material features. All this is reasonable and explainable. On the one hand, there are some products that have a direct relationship with the user, such as smartphones, which can be designed to be fully "perceived" by the consumer; in such cases the R-factor is relevant and more adjustable with appropriate choices for specific key-features. On the other hand, there are several other types of products that do not have such a strong relationship with the consumers' perceptions, such as, for example, the crankshaft in our classroom case study. Going into depth, as regards this case study we identified only three R-type key-features: the "rate of nonconformities," the "process product time and/or the number of process steps," and the "surface hardening treatment cycle time." Why do we identify these features as belonging to the R-type category? We have all the information necessary to reply, and we just need to remember what has already been said about "customer" definition. In Chap. 2, in looking at the basics of the generic QFD matrix applied to products (or services if you want), we learned that the customer is any party that is involved in the fruition of a product, not just its final user. In Chap. 3, we specifically talked about the product value chain and we said that if we design and sell bike saddles, the chain of our customers would be the bike manufacturer, the bike shops, the online after-market sellers and the biker. Now return to the previous question—the 3 material key-features that impact on one customer of our product value chain. Although such features do not have any relationship with the final user, who could be the car manufacturer, they can impact on the work of a sub-supplier, our customer, of the car manufacturer that buys the finished crankshaft from us to assemble with some other parts in motor engines. As likely as not, the assembler would have an internal organization that focuses on a high reduction in his process inventory, so any external factors that can help to reduce the cycle time of semifinished products insourced is judged positive. Choosing a material that requires specific treatment with a compact and constant cycle, and with low uncertainties, for example, could have a positive effect on our customers' receptiveness for our product because it favors that customer's work organization that helps to reduce the balance of the workstation inventory capacity, thus helping to insource the right quantity of materials in short times.

Now we can proceed with the last step. We use again the data from the *QFD4Mat* in Fig. 3.12. The data at the "foot" of the matrix are the results of the assessment of material key-features for each candidate material. We can reorganize them by

Table 4.2 Data extracted from the *QFD4Mat* of Fig. 3.12: the data at the "foot" of matrix, the assessment of material key-features for each candidate material have been reorganized by grouping them into the P, C, and R category identified in the "top roof" of the matrix (see again Fig. 4.13, boxes linked by "B") to calculate the mean values for the P-type, C-type, and R-type material key-features

Materials	P	C	R
UNI EN 18 NiCrMo 5—Carburized	3.2	3.5	3.0
UNI EN 42CrMo4—Induction hardened	3.5	3.7	4.7
UNI EN 42CrMo4—Nitrided	3.7	3.0	3.3
Ductile cast iron—induction hardened	2.3	4.6	3.7

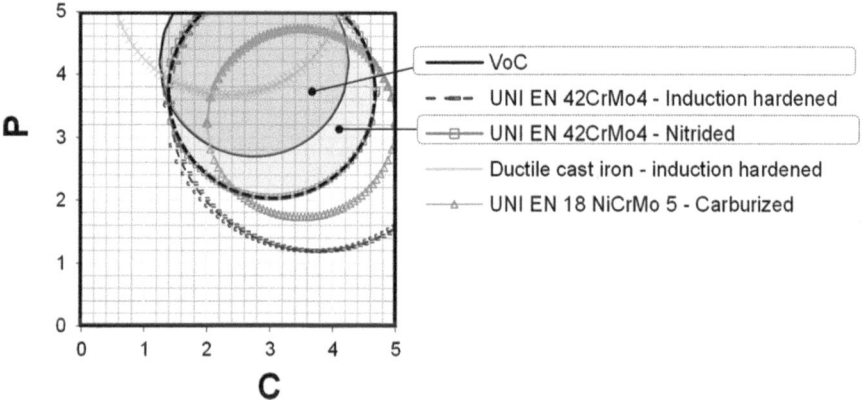

Fig. 4.14 The *VOC* and the candidate materials relative positioning in the PCR bubble map graph. The quenched and tempered steel treated by surface gas nitriding is the best choice for *VOC* matching among the candidates

grouping them into the *P, C,* or *R* category that is identified for each key-features at the "top roof" of the matrix (see again in Fig. 4.13, the two boxes linked by the letter "B"). By calculating the mean values for the P-type, C-type, and R-type material key-features, we obtained Table 4.2.

Thus, as we have already done for the *VOC* data, also the key-features (which, remember, represent the technical translation of the *VOC* requirements) can be summarized for each material in the three parameters P, C, and R.

Ultimately, from these three parameters we can plot the material's *key-features bubble* into the *PCR bubbles map*. The result we expect is something similar to what we discussed in general in the previous paragraph. Let us see if it fits. The results for the classroom case study are shown in Fig. 4.14.

So let the tool you downloaded from the Web site, which you can customize to your needs, do the job for you, and enjoy the improvement in your material research—or its supreme innovation in a revolutionary way.

Postface

There is a famous joke about how product development is generally managed in enterprises. It is a story about how an R&D manager, a marketing manager, and a sales manager decide to cooperate when they go hunting. The marketing manager convinces the sales manager that the real essence of hunting is to catch something big; the sales manager thus decides to go off alone into the savanna, and, having caught the attention of the lion, runs away toward the R&D manager telling him: "I've got its attention! Now you can catch it.". This is about the lack of cooperation, or more of a competitive instead of a collaborative approach, which should develop last during the work on a multidisciplinary project, as in product development processes. The master idea contained in this book is that no point of view ever has minor importance in multidisciplinary teams. There can only be an insufficient way of sharing disciplines, because in such cases there are so many incomprehensible languages, as in a board game with common rules for all the players, even when they have different strategies or scopes. The approach described in this book aims to communicate possible common rules, so aiding the single player, though he or she may have a genius for approaching a problem from his/her point of view, to put his or her real talent to use for the team as a whole.

A Few years ago, during technical meetings with company's managers, I was struggling to make out how is it that they did not understand what I was trying to explain about the technical aspects of the materials to be selected for product improvement. I thought, they were not really interested. One day, one of these managers decided to bear with the impatience and impertinence of a young researcher, and told me a story. The story took place in the early years of twentieth century, in North America. An inventor movedfrom Japan to the USA to study what was then a promising and new industry that was growing in the field of car manufacturing, with, as its main player, the Ford motor company. He took 4 months to comprehend the new sector (imagine never having driven an automobile, and never having even owned one). He observed the emerging American car manufacturing industry and wrote some notes about the huge potentialities of Ford, but he also took some notes about what he foresaw about the future limitations that he thought the American car

© Springer International Publishing Switzerland 2015

F. D'Errico, *Material Selections by a Hybrid Multi-Criteria Approach*,
SpringerBriefs in Materials, DOI 10.1007/978-3-319-13030-9

manufacturer would have faced in mature markets. Coming back to Japan he was convinced that the World was approaching the Automobile Era, and started to analyze how he would compete with the giant Ford motor company. His son, Kiichiro Toyoda, some years later decided to follow his father's insight. Using his great experience of another industry sector, he took on car manufacturing and founded the Toyota Motor Company and introduced one distinguishing criteria from the giant Ford—how to reduce costs by decreasing "losses" over the supply chain in order to produce many different cars for the many Japanese' tastes (Ohno 1998). The story of the slow economic growth of today is not so different from the many difficulties companies have experienced in the past; during slow growth economy periods and—more so than in the past—in the context of highly competitive mature markets. On account of our highly demanding and well-informed consumers, a great barrier to company growth is represented by a periods of overproduction or production with "product value losses." The value of a product needs to be created (Kim et al. 1997), that is reasonable; but we also need to do the job by using all our intelligence and all the resources we have in each room of our enterprise. I have tried just to give you a possible board game, to reduce as far as possible the effort you need to make in creating complicated and boring calculation tables and graphic analysis. Those tools can be customized, as you see fit from *www.qfd4mat.com*, so that "all" you need to do is to put your bright ideas on the table, and let your team to help you in composing your final successful puzzle.

References

Akao, Y. Quality Function Deployment: Integrating Customer Requirements into Product Design. Cambridge, MA:Productivity Press (1990)

Ashby, M.F. Materials selection in mechanical design (2nd ed), Butterworth Heinemann, Oxford, UK (1999)

Ashby, M.F. Materials Selection in Mechanical Design, Third Edition, Butterworth Heinemann, Oxford, (2005).

Ashby, M.F.; Cebon, D. Case studies in materials selection. Granta Design Ltd, Cambridge, UK (1999)

Ashby, M.F.; Johnson, K.; Materials and Design, the Art and Science of Materials Selection in Product Design Butterworth Heinemann, Oxford (2002)

Ashby, M.F.; Brechet, Y.J.M.; Cebon, D.; Salvo, L. Selection strategies for materials and processes. Materials and Design, 25 (2004)

Booz, Allen and Hamilton, New Products Management for the 1980s, Booz, Allen and Hamilton, New York, NY (1982)

Charles, J.A.; Crane, F.A.A.; Furness, J.A.G. Selection and use of engineering materials (3rd ed) Butterworth Heinemann Oxford, UK (1997)

Cobb, C. W.; Douglas, P. H. A theory of production. American Economic Review 18(1):139–165. Supplement, Papers and Proceedings of the Fortieth Annual Meeting of the American Economic Association (1928)

Cooper, R.G. Third generation new product processes, Journal of Product Innovation Management, Vol. 11, pp. 3–14 (1994)

Dieter, G.E. Engineering design, a materials and processing approach (2nd ed) McGraw-Hill, New York, USA (1991)

Farag, M.M. Selection of materials and manufacturing processes for engineering design, Prentice-Hall, Englewood Cliffs, NJ, USA (1989)

Gros, J. Progress through product language. Innovation: The Journal of the Industrial Design Society of America, 3 (2), 10–11. (1984)

Kim, W. C.; Mauborgne, R. Value innovation: The strategic logic of high growth. Harvard Business Review (January-February) (1997)

Kim, W.C.; Mauborgne, R. Blue Ocean Strategy: How to Create Uncontested Market Space and Make the Competition Irrelevant. Boston: Harvard Business School Press (2005)

Lewis, G. Selection of engineering materials, Prentice-Hall, Englewood Cliffs, NJ, USA (1990)

McKinsey, Energy efficiency: a compelling global resource, Sustainability & Resource productivity, 2010, available at: www.mckinsey.com (2010)

Mital, A.; Desai, A.; Subramanian, A.; Mital, A. Product Development: A Structured Approach to Consumer Product Development, Design, and Manufacture, second edition, Elsevier, 2014

© Springer International Publishing Switzerland 2015

F. D'Errico, *Material Selections by a Hybrid Multi-Criteria Approach*,
SpringerBriefs in Materials, DOI 10.1007/978-3-319-13030-9

121

Mowery, D.C.; Rosenberg, N. The influence of market demand upon innovation: a critical review of some recent empirical studies. Research Policy, Vol. 8 No. 2, pp. 103–53 (1978)

Ohno, T. Toyota Production System: Beyond Large-Scale Production, Productivity Press (1988)

Porter, M. E. Competitive Strategy: Techniques for Analyzing Industries and Competitors. New York: Free Press, 1980. (Republished with a new introduction, 1998)

Porter, M. E. On Competition. Updated and Expanded Ed. Boston: Harvard Business School Publishing, 2008

Rogers, E.M. Diffusion of Innovations, Fifth Edition, Free Press, New York, (2003)

Rothwell, R.; Soete, L. Technology and Economic Change. Physics in Technology 14 (6) November: 270–7 (1983)

Ulrich, Karl T.; Eppinger, Steven D. Product Design and Development, 5th ed. New York: McGraw-Hill Higher Education (2011)